PLANT BIOLOGY

PLANT BIOLOGY

AN OUTLINE OF THE PRINCIPLES
UNDERLYING PLANT ACTIVITY
AND STRUCTURE

BY

H. GODWIN, Sc.D., F.R.S.

**FELLOW OF CLARE COLLEGE, CAMBRIDGE,
AND UNIVERSITY READER IN
QUATERNARY RESEARCH**

CAMBRIDGE

AT THE UNIVERSITY PRESS

1959

CAMBRIDGE
UNIVERSITY PRESS

University Printing House, Cambridge CB2 8BS, United Kingdom

Cambridge University Press is part of the University of Cambridge.

It furthers the University's mission by disseminating knowledge in the pursuit of education, learning and research at the highest international levels of excellence.

www.cambridge.org
Information on this title: www.cambridge.org/9781107586437

© Cambridge University Press 1959

First edition 1930
Second edition 1933
Third edition, revised 1939
Reprinted 1943
Fourth edition 1945
Reprinted 1946, 1948, 1954, 1959
First paperback edition 2015

A catalogue record for this publication is available from the British Library

ISBN 978-1-107-58643-7 Paperback

NOTE ON THE THIRD EDITION

Since this book was first written the Universities of Oxford, Cambridge and London have combined, in the interests of schools, to make uniform the scope of teaching for the biology of their respective 1st M.B. examinations. A more or less common basis for this has been agreed upon, although the different universities themselves choose to spend different periods of time in teaching it. These changes have made it necessary to add to the book a consideration of the fern plant, and of reproduction by means of the flower and the formation of seeds. This has involved some mention of alternation of generations and reduction-division. *Pythium*, *Pellia* and *Sphaerella* have also been added to the lower types described, with the hope that the usefulness of the book will be increased to students taking other biology examinations, such as those for the Higher School Certificate. For suggestions for improving and correcting the text I have to thank many good friends, especially school teachers and students. In particular I am grateful to Dr C. S. Hanes who has lent his comprehensive knowledge of plant biochemistry to correcting the section on organic substances and their properties.

H. G.

October, 1938

NOTE ON THE FOURTH EDITION

This edition is substantially identical with the third edition, but recent additions to our knowledge have necessitated minor alterations. The opportunity has been taken to insert a note upon the nature of *Penicillin*, and to revise the very brief comment on the nature of meiosis.

H. G.

PREFACE

Text-books on animal biology are already fairly numerous but those devoted to plant biology are still so few that no excuse is needed in offering this new volume. Especially is this so since text-books, both of plant and animal biology, have hitherto been concerned most largely with the morphological and evolutionary aspects of the science, whilst the present volume, although not neglecting these, follows an increasing tendency of biological thought in laying greater emphasis on the physiological point of view and on consideration of the simpler characters of the physico-chemical background of plant-life. While based on elementary lectures given to first year medical students and designed primarily for their use, the book is intended to have also a wide utilisation by other biological students of similar status. It should be suitable for use in the higher forms of those schools in which biology is taught and in the introductory classes of training colleges. The experience of the writer as a member of the staff of the Cambridge Botany School confirms the opinion that too often students of botany not only come up to the University, but pass through its botanical courses with far too little appreciation of the general biological significance of the subjects with which they deal. It is hoped that this book will afford a means whereby such a deficiency may be made good.

The nature and scope of the subjects which form part of the training of the medical student are matters which for a long time have been debated by the medical and other University teachers responsible for them. It is of course desirable that the student should gain knowledge of some subject as a pure science, so that he may bring exact methods and exact thinking to his later work. Such would no doubt be valuable in each and every subject which he reads, but the field of medical practice and theory is so rapidly widening that there exists a natural and strong tendency to cut down preliminary introductory study to a minimum, and to excise all matter save that with direct and obvious application to later and more specialised medical work. Different universities vary in the extent to which they think such tendency should be yielded to, but as they all must find a practical solution, some kind of compromise usually follows. This takes the form of utilising in preliminary science courses matter as closely linked with later work as possible, and discarding things which may be part of the traditional teaching of the subject but which the medical student is not likely to meet with again. Such has been the background of this text-book; it has been designed largely to comprise material or principles which must later be more fully developed by the student. To this end considerable attention has been paid to the simpler aspects of physiology of the green plant and of the bacteria and fungi; the chemistry of simple organic substances and their occurrence in the plant in a colloidal state have been dealt with, thus creating a general background for later work on animal physiology, and illustrating the significance of general physiological treatment in the widest possible way. The bacteria and fungi, as organisms later subject to special study, have also received an attention which considers their biology as living organisms as well as their significance as disease producers, etc., to the rest of the world. In pursuance of such aims the study of the flowering plant has been reduced to an extent far beyond what is customary. Its

reproduction has been omitted and all details of structure
and anatomy save of such simple kind as will permit the
comprehension of the outlines of the chief physiological processes in the plant, and an understanding of the specialisation and differentiation of the land plant as the end
product of an evolutionary series of green plant organisms.
A certain number of plant types have been considered in
Chapters IX, X, XI and XII as illustrative of such evolutionary sequence, and by means of them have been expressed
such wide principles as differentiation of tissues and organs,
specialisation and division of labour, the origin of sex and
of a mortal plant body. Mention of alternation of generations has been omitted because, unless the principle is later
to be applied in detail to the most complex plant types, the
gain is not worth the considerable time necessary to teach
it. It is the hope of the writer that in the treatment of the
subject, little matter has been employed which will really be
useless once the medical student is beyond his first M.B.
examination, whilst at the same time no very important
principles shall have been neglected.

Whilst most of the text-book is intended to give the
student direct access to biological fact and outlook, its
limited size and wide scope together have made the book
one which, in the author's view, is well suited to expansion
and illustration by teacher or lecturer. This applies especially
to the parts of the book which deal with what we may call
the chemical background of plant physiology. It would
have been impossible to devote space here to the closer
explanation and the fuller treatment which the subject
should have: at the same time it would have been unwise
to omit all consideration of the chemical background
of plant physiology, for this type of knowledge surely

is basic biological fact of the widest possible value, and subject to application and development at many stages beyond the elementary examination in biology. It is difficult to develop the outlook here so briefly indicated, because traditional methods have ill prepared students for it, but it is to be hoped that this difficulty will diminish as teachers increasingly realise the importance of developing the theme, and as the theme comes to be borne in mind throughout both physico-chemical and biological teaching.

As the book has not been written for use by students of less than sixteen or seventeen years of age, it has seemed permissible to assume in the reader some elementary knowledge of physics and chemistry, but at the same time every attempt has been made to make the writing self-explanatory or to leave it in such a state that reference to a text-book of chemistry or physics will at once make clear unexplained parts. The importance to the biologist of an adequate training in chemistry and physics can hardly be over-estimated, and the medical man, whose work has well been called "applied biology," will find it, through both training and practice, to be of the greatest value.

It will be clear that although this book has been written without instructions as to possible practical work, it is intended that such work should accompany study of each part of the book as it accompanies the lecture course in Cambridge. Only by personal practical work can the student obtain the scientific habits of accuracy in observation and statement, which these courses are designed to promote.

The illustrations of the book have in most cases been redrawn and rearranged from figures in well-known botanical text-books; in such cases the source of the drawing is acknowledged in the subscription; others of the drawings are original. In the execution of all of them and in the preparation of the text, I have had the assistance of my wife, whom I should like now to thank. My gratitude is due to many others who have helped directly or indirectly in the production of this book; especially I owe much to Professor A. G. Tansley, Sherardian Professor of Botany in the University of Oxford, who, whilst lecturer in Cambridge, had much to do with framing the syllabus for the First M.B. biology, and in setting it on sound scientific lines. To him and to other members of the Botany School staff I owe sincerest thanks for the influence which, as teachers and friends, they cannot fail to have exerted on the preparation of this book. I am especially indebted to Mr F. T. Brooks, who has read through Chapter VIII, on the Fungi, and to Dr F. F. Blackman, who has beneficially criticised the biochemical and physiological parts of the book. My old friend and teacher, Mr S. Clegg, has most kindly helped in the correction of the proofs.

H. G.

December, 1929

NOTE ON THE SECOND EDITION

For the second edition some small amendments have been made. The newly observed facts of the conjugation of *Spirogyra*, of gametal release in *Fucus*, and of the hormone nature of the stimulus conduction in tropic responses have been mentioned; the section on bacterial structure and diagnosis has been altered to conform more closely to modern bacteriological practice and theory. For his kindness in giving advice on this last section I am indebted to Mr E. T. C. Spooner of the Department of Pathology, Cambridge.

H. G.

October, 1932

CONTENTS

CHAPTER I

THE LIVING PLANT

Living and non-living matter. Since biology is the science of living organisms, it is appropriate that a biological text-book such as this should indicate in the first place the characteristics which define living organisms, and since it deals with plant biology, it should also give some of the distinctions between animals and plants, and a general statement of the position of plants in the organic life of the world. These subjects will be the basis of Chapter 1.

The distinction between living and dead matter is a less simple matter than appears at first sight; we agree that a crystal of sodium chloride is dead, and an active human being is living, but the criteria on which the distinction is based remain undefined, and certain border-line cases are hard to resolve. Thus the muscles of the leg of a frog can be dissected out of the animal, and may be kept in Ringer's solution for some days still capable of normal contractile movements when electrically stimulated. The frog as a unit may be dead, but are we still entitled to regard the leg muscles themselves as dead? We can only approach such a question by examination of the distinctive characters of living and dead things.

The capability of movement is possibly the most obvious character of living things, but although evident in animals, the ability in plants is reduced to little more than growth movements far too slow to be seen by eye; movements of at least equal magnitude seem to be involved in the flow of rivers, convection currents, tides and so forth. Thus as an absolute and simple criterion, movement is eliminated.

It may be suggested that living things *grow*, but this in itself is an insufficient standard, for salt crystals, recognisably non-living, will grow when placed in supersaturated solutions. Nevertheless organic growth has certain qualities not found in crystal growth. Thus the organism takes into its body diverse chemical matter and *assimilates* it, that is builds it up into its own living material, protoplasm, so that this definite substance is always increasing by the intake of substances unlike itself, substances on other grounds regarded as dead. This growth, unlike crystal growth, is not indefinite but seems always directed to a definite end; when the organism reaches a certain adult size, it *reproduces* and gives rise to new individuals like the parent one but smaller. The apparent direction of the activities of the organism to this end seems a valid distinction between living and dead matter. Another such concerns *response to stimuli*. The response of a man to a pinprick by rapid movement can be paralleled by the slower response of plant stems in a dark room turning towards a lighted window, but the excised frog muscle still responsive to stimulus makes one question the fundamental, if not the general value of the criterion. The distinction between living and dead matter is most strikingly evident in the phenomenon of death, which marks a sharp cessation of all the activities we most commonly associate with living things. Such considerations as these convince us of the practicability of separating living and dead things for general purposes, but it is questionable whether such distinctions can be fundamental.

The claim of plants to be considered, equally with animals, as living organisms, must be evident from this brief discussion, but in Chapter IV a more detailed account of the various plant activities will be found, such as to leave no doubt that plants really do come into this category.

The uses of plants. The use of plants to the world at large is a matter which can only be properly estimated after a full knowledge has been obtained of the complex activities and complex interrelations of plant and animal life on the globe, but some points may be made now to be supplemented by others given in the later sections of this chapter and this book.

Plants form the basis of life on the globe because they possess the power of building up organic food, in large quantities, out of inorganic food. They thus supply food for all the animals living in the world and needing organic matter as food. Similarly they are the chief source of fuel in the world, whether coal, peat, lignite, wood or oil. They afford a ready though not the only source of vitamines, and since they include the bacteria and fungi, they must be considered as responsible for plant and animal diseases, for various fermentations and putrefactive actions which break down dead organic matter and animal excretions, which might otherwise, by their accumulation, put an end to animal life on the earth. Further extension and application of such uses will be found throughout several of the ensuing chapters of the book.

Plants and animals. There is little real difficulty in distinguishing between plants and animals; the highest types on each side are readily recognisable. Animals move about freely and plants do not, plants are commonly much branched and animals compact, and of the two groups, plants alone are predominantly green. These are perhaps the most striking differences and those most evident in a superficial comparison between two such organisms as a man and an oak tree. Yet to these differences of movement, shape and colour must be added differences in the mode of nutrition, of cellular structure and growth. The most fundamental differences of all lie in the nutrition, and to this all the other differences may be easily related. The nutrition of the green plant depends on the possession of *chlorophyll*, a complex mixture of green and yellow pigments borne on the surface of small protein bodies, called *chloroplasts*, which are found in immense numbers in every green leaf and which, in some form or other, are present in all green plants. These bodies, when exposed to light, are able by absorbing the radiant energy to convert carbon dioxide into simple organic material such as sugar. This is the process known as *photosynthesis*. Supplies of carbon dioxide constantly diffuse from the air towards these places where it is being used up, and so the green parts of a plant are able to synthesise large amounts of sugar which form the starting point of the synthesis of all the complex organic material of the plant body itself. The plant nutrition also involves the intake of mineral substances in solution from the soil, their combination with the substances produced in photo-synthesis, and the synthesis of both into still more complex organic material. Thus the plant food is inorganic, and the plant synthesises organic material from it. Such a function is quite lacking in the animal world, and it is upon this difference that all the chief distinctions between animals and plants eventually rest. The possession of chlorophyll is the key to *autotrophic* existence, that is nourishment by a system which is quite independent of other organisms. Because they have chlorophyll, the plants could readily live alone in the world, but the animals could not exist without the plant life which constitutes their food supply. As the animal eats solid food so it must have a mouth or gullet, a digestive region in which this food is rendered suitable for utilisation by the body, and excretory mechanisms by which residue can be disposed of. None of these exists in the plant, but the green colour of the plant indicates its photosynthetic powers, and the branched and spreading form of leaves and branches is significant as offering an extremely large surface for the absorption of inorganic substances present in great dilution in the soil or in the air. The widely spread leaves are significant also as an efficient means of absorbing light.

Since the animal cannot make organic material it must live on organic food made by plants, and to seek its plant food it must be able to move about. Thus its powers of locomotion and its compact form are seen to be implicit in its mode of nutrition. The differences between plants and animals in cellular structure and in localisation of growth will be dealt with later, and they too will be found to be very obviously related to the nutritional processes.

From this it will be clear that the safest single criterion for distinguishing a plant from an animal will be the possession of the chlorophyll pigment. All the other distinctions are vague and exceptions can be found to them.

Thus there are fixed animals such as sea-anemones and sponges. There are branched animals such as the corals and compact plants such as the cacti. Finally there are, in the sea, in river water and in rain water, simple organisms hard to classify either as plants or as animals. According to the theory of evolution all species of living organism have developed by a process of gradual change from a few simple original types, and there exist to-day a few simple organisms probably much like these ancient types. For example, there are tiny green unicellular organisms which absorb their food in liquid form but which are also motile like a true animal. Somewhere in ancient history, the main separations have been made from types like these into animal and plant lines of evolution. Indeed such separation may have occurred repeatedly, but in any case from such a point the organic world has diverged into the holozoic and the holophytic classes—that is, those classes entirely animal, and those entirely plant-like in their nutrition—the holozoic, dependent and the holophytic, independent.

The comparisons outlined so far between plants and animals, though sufficiently striking, can be supplemented by a view at a different angle, of the relation to each other forced upon them by their different modes of life. And in this case the dependence of the human race itself upon the plant world may be more fully recognised. This view involves a consideration of the energy relationships of animals and plants. In both animals and plants, by the process of *respiration*, compounds of high energy content are oxidised to compounds of lower energy content, and the energy set free produces heat and movement and maintains the various activities of the life of the organism. Perhaps the greatest simplification of this idea is to say that the following equation represents the respiration process in both plant and animal cells.

$$C_6H_{12}O_6 + 6O_2 \longrightarrow 6CO_2 + 6H_2O + 677 \text{ kg. calories}[1].$$
(glucose) (oxygen)   (carbon   (water)
                            dioxide)

This process going on in all living cells supplies energy for growth and movement and maintenance. If such a consumption of organic material of high energy content is constantly going on by both plants and animals it might be expected that shortly all such organic matter in the world would be consumed, and so indeed it would, save for the fact that plants can synthesise this organic material, by means of the chlorophyll in the cells of the leaves. This

---

[1] 1 kg. calorie, i.e. 1 kilogram calorie, is the amount of heat required to raise the temperature of 1 kilogram of water through 1° C. The equation indicates that 180 grams of glucose (1 gram molecule) on complete oxidation liberates 677 kg. calories.

can be empirically stated as the reverse of the former
equation:

$$6CO_2 + 6H_2O + 677 \text{ kg. calories} \rightarrow C_6H_{12}O_6 + 6O_2.$$

As a result of this process of *photosynthesis* or *carbon-assimilation*, carbohydrates are produced by the plant which
have a greater energy content than the equivalent carbon
dioxide from which they are formed.

This plant process is the chief process by means of which
new energy is fixed on the planet. The winds and river-flow are due to sun energy, and the tides are lunar, but all
our common fuels such as coal, peat, petrol and wood burn
with energy derived from sunlight, by photosynthesis.
George Stephenson is said to have explained enthusiasti-cally that it was sunlight stored millions of years in coal which
enabled his locomotive engines to pull their trains along.

Thus the supply of fuel which is oxidised to give power
and heat depends closely on photosynthesis, but the direct
use of plant products as fuel is less important than the other
slower process of respiration which, as we have said, is also
an oxidation process going on slowly, but continually and
essentially, in all plant and animal life. In this also the
energy release is the essential part of the reaction, and only
by the photosynthesis of plants are supplies of high energy
organic material produced.

In plants both photosynthesis and respiration go on, and
the former predominates, so that not only can a plant
produce all the organic material which it can respire, but
it has an overplus which is stored in various ways, for
example in roots, stems, leaves, cell-walls, seeds, etc.

It is by consuming plants that animals acquire the
energised products for their respiration, or by eating animals
which have previously eaten plants. The energy produced
may result in movement or in heat or it may be used to

synthesise living material of even greater energy content. Thus one animal is effective food for another, but no animal has the capacity for utilising inorganic substances and radiant solar energy on a large scale and thereby becoming independent. For as it is, all life is dependent upon green plants and this is the deeper meaning of the phrase "all flesh is grass." Even man is bound by the same ties, though the progress of research may in time alter this. Should photosynthesis, the utilisation of solar energy in the formation of sugar from carbon dioxide, ever become possible *in vitro*, we may envisage an age of food and fuel and power as cheap as the town water supply. That day is not yet, but the not too visionary picture will shew the unrealised significance of photosynthesis and the plant world to mankind.

Cellular structure of plants. Plants and animals are made of protoplasm and its products. Whilst protoplasm itself is the fundamental living matter of the organism the great bulk of a complex plant body is made up of the non-living substances secreted as a kind of skeleton by the protoplasm. The protoplasm is split up into more or less separate units called cells and each of these secretes a cell-wall round it. This wall behaves as a sort of external skeleton—an exoskeleton, like that of a crab or lobster—and upon it devolve practically all the mechanical functions of the living plant. This theory of the cellular composition of plants and animals was propounded by Schleiden and Schwann in 1831 and now appears almost as a self-evident fact.

Each cell has a "nucleus," and in a young state each cell is capable of division and extension almost indefinitely, but all the same, a multicellular plant is much more than a collection of living units in a common skin. The protoplasm of each living cell extends through tiny perforations in the walls into the neighbouring cells so that there is complete protoplasmic continuity from end to end of the plant. The demonstration of this fact made a great change in the general conception of plant structure and plant physiology. For instance, this discovery provided some path along which might pass the conduction of stimuli from cell to cell of the plant, a process recognisable in plants just as in animals despite the absence of any definite nervous system. These very fine filaments of protoplasm explained, or rather their presence agreed readily with, the behaviour of the plant as a single organism, rather than as a complex colony of micro-organisms.

The plant cell is commonly a box-like body about $\frac{1}{20}$ of a millimetre long. The protoplasm lines the inside as a colloidal layer. The wall itself is largely cellulose and may in some cases be observed to be laid down in layers from the inside by the protoplasm. The space inside the proto-plasm is called the vacuole and contains "cell-sap," a fluid which is a dilute solution of organic and inorganic substances, mostly useful as material or secreted as products of the cell's activities. The outer layer of the protoplasm, next to the cell-wall and next to the vacuole, forms a structure which from its physico-chemical characteristics is of the greatest importance to the life of the cell. It cannot be distinguished as a separate layer under the microscope.

The protoplasmic layer contains the *nucleus*, the most complex unit in the plant cell, and as in the animal cell, a body of the utmost significance in the reproduction and multiplication of the cell. The nucleus can be identified as a definite, almost spherical, granular body (Fig. 1). Other smaller structures called plastids occur in the protoplasm and they may be coloured green by the presence of a surface layer of chlorophyll, in which case they are called chloro-plasts. They are generally ovoid or lens-shaped masses of protoplasm. In the layer of chlorophyll on the surface of these plastids, photosynthesis takes place, and the absorption of carbon dioxide, the formation of carbohydrates and the giving out of oxygen involved in this process.

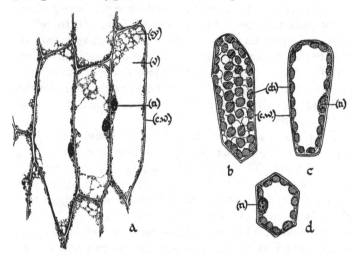

Fig. 1. Living cells. *a*, cells from the skin of the fleshy leaves of the Onion bulb; they have thin cellulose walls (*c.w.*), lined with a thin layer of cytoplasm (*cy*), which contains the ovoid nucleus (*n*). The greatest part of the interior of the cell is occupied by the vacuole (*v*) containing cell-sap. *b*, *c*, *d*, green photosynthetic cells from the Iris leaf containing abundant chloroplasts (*ch*). *b*, cell seen in side-view shewing chloroplasts on all the walls; *c*, cell in longitudinal section shewing the nucleus (*n*) and chloroplasts in the cytoplasm lining the cellulose wall (*c.w.*); *d*, the same cell seen in transverse section.

The small size of the plastids gives an enormous surface area for the bulk of plastid material, so great that the chloroplast surface in a big leaf of a castor oil plant would be over a hundred square yards. The chlorophyll can be extracted from the surface of these bodies but will not then photosynthesise. The plastids may be colourless, as in tubers for instance, and then they are called leucoplasts. They serve, in such cases, as centres of starch formation.

Not all the plant body is made up of living plant cells

save in the smaller and simpler plants. In a large tree such
as an oak, the bulk is formed of the dead skeletons of cells
which have lost their living contents. The tree relies considerably on the support and food-carrying capacity of
these dead cells. This state of affairs is not common in
the higher animals. We shall have to deal to some extent
with these structures later. For the moment we are concerned with the living protoplasmic structure of the cell
and the physico-chemical properties on which its manifestations depend.

A short consideration of the structure of one or two very
simple organisms may serve at this stage to illustrate both
the characteristics of cell structure and the characters of
organisms which are not very definitely plant or animal.

Chlamydomonas. This may be considered as the simplest
type of plant organism. It occurs abundantly in rain-water
butts and similar places. It consists of a single microscopic
ovoid or oblong cell from one end of which project two long
thin threads of protoplasm, the cilia or flagella, which by a
rapid wave movement passing from the base towards the
free end of the cilium, propel the cell swiftly through the
water, the cell rotating meanwhile about its long axis (Fig. 2).
A thick cellulose wall encloses the protoplasm and it thickens
out at the forward end from which the cilia project. At this
end also, just within the wall and in close contact with the
cilia, are two "contractile vacuoles" which first expand and
then suddenly discharge their contents to the exterior of
the cell, contracting as they do so. The two vacuoles
alternate with one another in this operation. Close to these,
and also just within the cell-wall, is a red "eye spot"
containing the pigment carotin. This eye spot is often
sensitive to light and is regarded as controlling in some
way the direction of movement of the *Chlamydomonas*.
However they "perceive," it is clear at least that the cells
do swim towards light of moderate intensity, so that some
kind of sensitiveness must be granted to them.

The very centre of the cell is occupied by the nucleus which is, however, ordinarily hidden by being within the

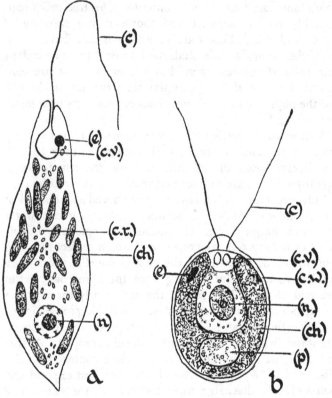

Fig. 2. *a, Euglena; b, Chlamydomonas.* The lettering for both is the same; (*c*), cilium; (*ch*), chloroplast; (*c.v.*), contractile vacuole; (*n*), nucleus; (*c.w.*), cell-wall; (*p*), pyrenoid; (*c.r.*), granules of carbohydrate. In *a*, the cell is seen in surface view; but in *b*, the cell is shewn in section to disclose the nucleus which would otherwise be hidden within the basin-shaped chloroplast. (*a*, after Doflein.)

deep basin-shaped chloroplast which occupies the greater part of the cell. A large bright body embedded in the chloroplast, which might at first be taken for the nucleus, is the "pyrenoid," a body probably concerned with starch formation. When growing in the open where light and carbon dioxide are available, the *Chlamydomonas* cell can live in a solution of weak inorganic salts which pass in through the cellulose walls in solution. A suitable salt solution would contain, for example,

Calcium nitrate   Magnesium sulphate
Potassium nitrate   Potassium phosphate
Ferric chloride.

Sugar and starch would be synthesised from the carbon dioxide and with the mineral salts the plant cell could survive and grow and reproduce. This is the nutrition of a plant and the only animal character present is the power of movement. We may at this stage recall the structure and nutrition of *Amoeba*. Here the protoplasm is naked and has no cellulose wall, and in consequence solid food particles are readily ingested by the organism and solid residues are expelled from it. There is no chlorophyll present, so that the source of energy used in movement and other vital activities must be the organic food material of plant or animal origin ingested in solid form. From the same source also must come all the materials requisite for the maintenance and building up of the protoplasm. Thus the organic materials taken in by the *Amoeba* are reduced to simpler form inside a vacuole in the body; that is they are "digested," a process developed greatly in higher animals. It has recently been shewn that amoebae are very active in ingesting and digesting bacteria, including, under laboratory conditions, many types pathogenic to man: in this way the amoebae resemble the phagocytes in the blood-stream. The absence of cell-wall permits changes of shape in the cell as external and internal conditions vary. Thus *Amoeba* shews a *holozoic* (completely animal) mode of life as contrasted with *Chlamydomonas*, which has a *holophytic* existence.

The flagellate, *Euglena* (Fig. 2), is difficult to classify either as animal or plant. It has one cilium which gives it mobility, it has no rigid cell-wall and so can readily change its shape. A large contractile vacuolar reservoir at the ciliated end of the cell communicates with the exterior by a channel. It is surrounded by smaller contractile vacuoles which, when they contract expel their contents into the reservoir. This in turn rythmically discharges to the exterior. The precise function of the contractile vacuoles in freshwater algal forms is not understood, although in simple animals they appear to be concerned with excretion. It is uncertain whether they are directly concerned with the intake of food substances. The cell of *Euglena* contains one or more chloroplasts, although when grown in solutions containing soluble organic food these may degenerate and finally disappear. Nevertheless, plants grown as saprophytes and devoid of chlorophyll for many generations retain the power to form green chloroplasts at once when replaced in dilute solutions of inorganic salts in the light once more. *Polytoma* (Fig. 22) is an organism like *Chlamydomonas* in all respects (including firm wall and corresponding inability to absorb solid food) save the absence of chlorophyll. It lives by absorbing soluble organic food material. Here, in types such as *Euglena* and *Polytoma*, where the final criterion, the nutrition itself, is a variable character, the organism cannot be said to be either plant or animal, nor is it necessary that it should. In no case, in biology, do the facts fit rigid rules—not because there are no rules, nor because the rules are necessarily different from the known ones which explain the behaviour of physicochemical systems, but because the deep complexity of living matter has never yet been met with sufficiently complex analysis and examination. The important thing to recognise at first is the existence of the *main types*, whether of function or structure, and accept for the time being less prominent divergences.

CHAPTER II

CRYSTALLOIDS AND COLLOIDS

Many of the substances of which the living plant is built are in the *colloidal* condition, and so some account must be given of this state of matter. The term *colloid* came into use about 1861, when Thomas Graham carried out certain experiments as a result of which he separated substances into crystalloids and colloids.

He subjected solutions of different substances to "dialysis" in a parchment vessel, or a vessel closed with parchment, which contained a solution of the substance under investigation, and which was suspended in a vessel of the pure solvent. Graham discovered that one class of substances passed readily through the membrane into the solvent, but the other class did not pass at all or passed only with extreme slowness. There was actually no definite discontinuity between the two classes of substances as can be seen in the table given below, which shews the times taken for equal quantities of different substances to diffuse through the parchment membrane, water being the solvent.

	Time units
Hydrochloric acid	1
Sodium chloride	2·5
Cane sugar	7
Egg albumen	49
Caramel	98

Graham noticed that the readily diffusible compounds were crystalline and the others non-crystalline. He therefore classified all substances as crystalloids and colloids. In this he was wrong. It has since been discovered that great numbers of typically crystalloid substances, such as sodium chloride, can be produced in a colloidal condition. Similarly silica, which might be taken, from its very abundant natural occurrence as quartz in granite, sandstone, etc., to be essentially a crystalloid substance. Thus the "colloidal condition" is to be spoken of rather than a "colloid kind of substance," and colloid and crystalloid are terms applicable to states of matter.

Many of the metals, commonly regarded as quite insoluble in water, can be prepared by various methods in a condition which at first appears to be that of a true solution. Gold, silver and platinum are all readily obtainable in this state, but the gold so obtained in solution will not dialyse. These solutions are spoken of as *sols*, or colloid solutions.

It is generally believed that such a solution consists of exceedingly small solid particles, suspended permanently in the solvent. The size of these particles, their nature, and the evidence as to their properties we shall consider later. The particles are spoken of as the *disperse* phase and the solvent as the *continuous* phase. The whole has the properties of a liquid. It is however really a mixture of solid particles in a liquid, i.e. the *disperse* phase is solid and the *continuous* phase liquid. This is called a *suspensoid*. Examples of suspensoids are given by all the metallic sols and by many water-colour pigments such as ultramarine, gamboge, etc.

The suspensoid is not far removed from the *suspension* which is, of course, the result when an insoluble but finely divided substance is shaken up with a liquid. The substance, for example sand, will settle sooner or later, or if lighter than the liquid would rise leaving a clear liquid below. With the colloid suspensoid the particles are apparently so fine that they stay indefinitely, and stably suspended, generally making the solution opalescent.

Now on drying or adding electrolytes these suspensoids (sols) may set to a solid form like a jelly. Such forms have the general properties of solids and are called *gels*.

In gels, the *disperse phase* is said to be droplets of liquid enmeshed in a solid framework which is the continuous phase.

The last state to be considered is the *emulsoid*. Here both phases are liquid, the emulsoid being to the simple oil-water *emulsion* just what the suspensoid is to the suspension. The particles or droplets are so finely divided as to be stable.

Just like some suspensoids, some emulsoids set to a *gel* state on standing or on cooling. Examples of commonly occurring suspensoids are starch solution, gum arabic solution, or protein. We imagine that a concentrated solution of protein exists in droplets, separated by a dilute solution of protein.

All the foregoing have been water-colloids, but not all colloids are of this nature. Great numbers are known to be formed with organic and other solvents, and many have considerable economic significance, for example, the cellulose sol made by dissolving cellulose in Schweitzer's reagent (copper oxide ammonia) and used in the manufacture of artificial silk.

We will consider again the *sol* type of colloid with a view to a closer examination of its properties and especially of the size of particles, which, we shall see later, is largely responsible for those properties.

Size of particles. In dialysis, small molecules, atoms and ions pass readily through the membranes of parchment, collodion or fish bladder. Colloids are kept back, and it seems reasonable to suppose that this is because the colloid particles, or aggregates of particles, are too large to pass through the membrane pores. The molecules themselves in some cases might be too large to pass. The dye "congo red," which is on the border-line between a colloid and a true solution, has a molecule of 72 atoms and a molecular weight of 654. Quite simple proteins have much more formidable molecules, e.g. albumin has been given the formula $C_{72}H_{112}N_{18}O_{22}S$ and the molecular weights of various proteins have been placed between 8000 and 200,000.

The soluble metallic sols, since the molecular weight is small, must certainly consist of aggregates of molecules.

By the process of *ultra-filtration* it is possible to get an idea of the size of the particle involved. This process involves the preparation of a graded series of collodion filters and the measurement by physical means of the size of their pores. These pores range in size from $930\mu\mu$ to 21μ ($\mu =$ one thousandth of a millimetre and $\mu\mu$ one millionth of a milli-metre). The pores range in diameter from about $\frac{1}{1000}$ mm. to $\frac{1}{50}$ mm. When samples of a given colloidal solution are placed in dialysing vessels made from such filters it will be found that the colloid will diffuse through all filters, the pores of which are above a certain size, and through none with pores below that size. In this manner a rough idea may be gained of the size of the particle of the colloid.

Although the colloid particles are too small to be seen through a microscope, their presence can be detected by the Tyndall beam effect. The basis of this method lies in the fact that when a beam of light is passed through a true solution it is not visible as a beam, but when passed through a colloid solution it forms a cloudy beam like a ray of sun-light in dusty air. This effect is due to the light reflected from the surface of the particles and it implies that the particles must be small compared with the wave-length of light, which is from $450–760\mu\mu$ for visible light. If a strong beam of light passing through a colloidal solution is viewed from the side through the high-power lenses of a microscope, although the colloid particles themselves remain too small to be seen, yet the presence of each one is detectable as a separate speck of light. The number of particles detectable as points of light can be counted in a given volume of solution of known strength and so the size of particle in any given sol can be approximately worked out. The apparatus used for this purpose is termed the ultra-microscope and with its aid, in sunlight, particles of $5\mu\mu$ can be detected. The values for sizes of particles obtained in this way agree well with those obtained by ultra-filtration. The ultra-microscope shews the particles always in violent motion and this movement is less in solutions in which the particles are of larger size. In a coarse suspension, such as sand, no movement of the particles is visible, but in a suspension of which the particles can be seen under the usual high power of the microscope, they are always found in violent and irregular movement. This phenomenon is called Brownian movement after Robert Brown, the botanist, by whom it was first observed. It can be readily seen by making up a solution of "burnt sienna" water-colour pigment and observing it under the high power of a microscope. It is now generally agreed that the movement is due to the unequal bombardment on the different sides of the particle by the molecules of the solvent. The larger the particles the less unequal would this bombardment be and the smaller the effect evident as Brownian movement. There seems no sharp break in this phenomenon between the smallest aggregate particles and the large molecules. A calculation from the displacement of particles in Brownian movement affords yet another method of determining the size of the particles in a colloidal solution. The result of measurements by all methods is to shew that solutions do not exhibit colloidal characters unless the particles in the solution are of a certain minimal size.

One immediate result of large molecules or large aggregates of molecules is seen in the osmotic pressure effects. Colloids give very slight osmotic pressures and also raise or lower the boiling and freezing points of solvents very little. The lowering of the freezing point (or raising of the boiling point) is, in any solution, proportional to the number of effective molecules, and in a colloid solution these are so much bigger than in a true solution that they are in far smaller numbers and yield correspondingly small effects. Though large compared with single molecules, colloid particles are extremely small when compared with visible solid particles, and the result of the fine subdivision of all the particles of a solid substance brings new factors into play of which the first is the enormously increased surface. If we start with a centimetre cube, with a surface of 6 sq. cms., and divide it into millimetre cubes, each has 6 sq. mm. surface and there are 1000 of them, so that the area is now 60 sq. cms., i.e. ten times the previous area. Thus the surface area increases directly as the linear dimension decreases. If a centimetre cube were cut up into cubes of ultra-microscopic size, i.e. $10\mu\mu$ (that is 1×10^{-6} cms.), they would have an aggregate surface of 6 sq. cms. $\times 10^6 = 600$ square metres. This area exposed by dividing a centimetre cube into units

Length of edges of particles	Specific surface	
1 cm.	6 sq. cms.	
1 mm.	60 "	
10μ	6,000 "	
6μ	10,000 "	(1 sq. metre)
1μ	60,000 "	
$100\mu\mu$	600,000 "	(60 sq. metres)
$10\mu\mu$	6,000,000 "	
$5\mu\mu$	Limit of ultra-microscope	
$2\cdot5\mu\mu$	Haemoglobin and fine sols	
$\cdot5\mu\mu$	Alcohol molecules	
$\cdot1\mu\mu$	Hydrogen ion	

of any given size is called the "specific surface" for that size of unit. Disperse systems begin to shew colloidal characters at a specific surface of 10^5 square centimetres (i.e. 10 square metres), though Brownian movement can be detected in particles of 10^4 square centimetres specific surface. The table on p. 20 illustrates the sizes of particles and their specific surfaces.

Colloid properties. The large surface exposed by colloidal particles is a very important factor in their physico-chemical properties. In those reactions in which platinum is employed as a catalyst, it is far more effective used in a colloid state than as the solid metal; a result probably connected with far greater active surface of the colloid. The great increase in surface has great effects upon surface energy. Surface tension is always tending to diminish the surface of a body to a minimum, as can be seen in the spherical condition assumed by drops of water when no external forces interfere. Consequently any expansion of the surface must be accompanied by work, and the surface, when produced, is the seat of energy. Roughly speaking one may picture the molecules dragged out farther apart into a condition of strain upon whatever linkages bind them. Under these conditions any foreign molecules settling on the surface would diminish the strain and lower the surface tension. There would thus be a marked tendency for the foreign substances to be *adsorbed* on such a surface. The colloids possess this capacity of adsorption to a marked degree. Solids as well as liquids of course shew surface tension though it cannot be directly measured. On account of this property china clay and fuller's earth, both natural substances found in an extremely fine state of division, are used as cleansing agents, and gelatine and isinglass are used to clear organic solutions such as beer.

Certain colloids shew an adsorption of certain substances

only (selective adsorption). Thus although iodine in sea water is so dilute as to be scarcely detectable, the selective adsorption of the colloids in seaweeds brings together such large quantities of this element, that the kelps are a profitable commercial source of iodine. Such selective colloidal adsorption by the protoplasm of the living cells may explain how the roots of crop plants can obtain from very dilute soil solutions such large quantities of mineral salts that their ultimate concentration in the plant may be relatively enormous. It has been ascertained that if a plant cell is allowed to stand in $N/5000$ calcium chloride, over fifteen times this concentration is achieved inside the cell before absorption ceases. It has also been shewn that when plants are grown in nutrient solutions, the concentration of the nutrients may be lowered from 2 per cent. to 01 per cent. without affecting the rate of growth of the cultures. Selective adsorption is very readily demonstrated by plants grown in water cultures, since in no case does the mineral composition of the adult plant, when analysed, correspond at all quantitatively with the mineral composition of the nutrient solution. There yet remain to be considered a few characteristic colloid properties. Gelatine is an example of a colloid which takes up water readily to form a sol and loses it to form a gel. The process, besides being completely reversible, is quite continuous and can be controlled by adding water or drying and also by heating or cooling. Because the change is continuous, it is possible to say that no true chemical compounds are being formed, such for example, as a definite gelatine hydrate. In the case of blue copper sulphate ($CuSO_4 5H_2O$), on heating the salt loses 4 molecules of water at $100°C$. and the 5th at $200°$, and the change at each temperature is a sharp one, corresponding with a definite change from one hydrate to another. In some cases gel formation is not reversible; this is the case in the *coagulation* which is produced by heating protein sols such as egg albumen; the reaction is not reversible and gives rise to a gel, but the properties of the protein are so much altered that some alteration in the protein molecules themselves, rather than in their aggregation, is supposed to have occurred. Similar coagulation is produced by some fixing agents such as alcohol and acetone. Another very important colloid character involves the rate of diffusion of substances through colloids. Whilst colloids themselves diffuse very slowly, probably by reason of their large molecules or molecule aggregates, salts will diffuse through them even when they are in the solid gel state almost as readily as through water. Most photographic processes are made possible by this fact. Developing, toning and fixing only follow the diffusion of the various solutions into the gelatine gel which is spread on celluloid or paper or glass, and which contains the sensitive silver salts. This rapidity of diffusion through all but the most concentrated gels suggests that the colloidal condition of the protoplasm will not in this respect diminish its activity.

The colloidal nature of the protoplasm. The protoplasm which lines the cellulose walls of the living plant cell is generally supposed to be in a colloid suspensoid or emulsoid state, in which the continuous phase is a dilute solution of various salts, sugars, amino-acids and other crystalloids, whilst the disperse phase is made up of protein particles or droplets. All such structure is invisible to the naked eye, being of ultra-microscopic size, but always within the protoplasm small microscopic granules or drops are visible and these do not form part of the essential permanent protoplasmic structure but are small aggregations of fats, carbohydrates or proteins which have been temporarily formed as a result of the protoplasmic activity. They may be referred to as "metabolites."

(a) *Adsorptive properties.* We have already illustrated the way in which the adsorptive properties of colloids may be held responsible for the taking up of large amounts of salts and sugars which continually diffuse into the cell. The staining of the protoplasm, both living and dead, by means of dyes is an illustration of the same phenomenon.

(b) *Coagulation.* The protein sols which form a large part of this mixed protoplasmic colloid are readily and irreversibly coagulated by high temperatures, and with this coagulation the life of the cell comes to an end. Similar coagulation can be produced by fixatives. This explains many of those methods of sterilisation which involve the killing of bacterial protoplasm either by heat or by antiseptics.

(c) *Sol-gel balance.* Within fairly wide limits the protoplasm can lose or gain water in the manner of gelatine, passing from one condition to the other readily and reversibly according to conditions. Roughly speaking we may say that at low temperatures ($0°$C.) the protoplasm tends to be in the gel condition and with higher temperatures to be in the sol state. Above $40°$C., however, the structure of the protein molecules is affected and the irreversible coagulation changes set in. Thus the protoplasm can be said to exist normally in a state between that of a sol and a gel, and a change in either direction is very readily caused by temperature, changed water content, changed salt concentrations and by different hydrogen-ion concentrations. It is very probable that many of the chemical activities going on in the protoplasm depend largely upon this constant flux in its colloidal nature and this colloid character clearly limits the temperature within which living organisms can exist and probably has some relation to the intolerance of desiccation shewn by most plant and animal tissues. Probably an effect of temperature change even more important to protoplasmic activity than the altered sol-gel balance, is that upon the reaction-velocity of chemical changes. The reaction-velocity of a chemical change *in vitro* is approximately doubled by each rise of 10° C., and there is every reason to think that chemical reactions *in vivo* behave similarly. The rates of the processes of photosynthesis and plant respiration have indeed been shewn to change with temperature in this manner, doubling for each rise of 10° C., until with temperatures of about 40° C., the killing of the protoplasm supervenes.

In certain cases, such as the imbibition of water by germinating seeds, the water uptake is due to colloid gels which may, as they swell, develop forces capable of splitting open the strongest of seed coverings. If dried peas are packed closely inside a glass bottle and water is added, the swelling of the peas will break the bottle. It is by similar forces that structures such as the hazel nut and plum stone are broken by the embryo plant.

(*d*) *Diffusion.* The case of diffusion through the protoplasm has already been mentioned, and it naturally provides a facility for reaction within it which is almost as great as if the medium were a liquid solvent such as water. In addition to these striking illustrations of the dependence of protoplasm upon colloidal characters, there are to be discussed some very important phenomena associated with colloidal membranes.

Osmosis and plasmolysis. In most cases a sol in contact with some other substance forms a gel membrane at the intersurface and this often has peculiar properties. This is indeed the case with the semi-sol, semi-gel colloid which represents the protoplasm in the cell, and gel membranes of peculiar properties are formed where the cytoplasm meets the vacuole inside and the wet cell-wall outside. If two solutions of different concentration are put into contact in a vessel, they will by the molecular movement of their constituents mix to form a uniform solution. Separated by such a colloid membrane as *parchment* the diffusion of the two solutions still goes on, but, as we have seen, whilst the small molecules of the solute and of "crystalloids" readily pass through the spaces, the larger colloidal particles are prevented. The solute molecules of a sol would readily pass the parchment. The parchment is said to be *permeable* to the solute and the membrane is said to be permeable without further qualification. However there are, as we said before, grades in the meshes of these colloidal sieves and parchment is about the coarsest. If we use a membrane such as pig's bladder or collodion, or better still a colloidal deposit (copper ferrocyanide) inside the pores of a porous pot, then the pores will be too small to allow any molecules to pass which are much larger than those of water. Thus to a sugar solution such a membrane would be *semi-permeable*. The water molecules would readily pass through the membrane but the sugar would not be able, and this is the circumstance which causes the phenomenon of *osmosis*. The usual experiment to demonstrate osmosis utilises a semi-permeable membrane, such as a parchment bag, containing a sugar solution, and suspended in a vessel of the pure solvent (water). Water passes into the sugar solution until the hydrostatic pressure of the raised column of solution is equal to the osmotic pressure of the solution at that time. (With a membrane so coarsely perforated as parchment, the sugar continuously diffuses through to the outer solution, but this is not sufficiently rapid to affect the demonstration of osmosis qualitatively.) The osmotic pressure is directly proportional to the temperature and to the concentration of the solute, and if two solutions of different strength were employed inside and outside the membrane the osmotic pressure would be the difference between the osmotic pressures of the two solutions. Owing to the semi-permeable character of the gel membranes round the protoplasm, or even of the protoplasm itself, we shall find that osmotic phenomena play an extremely important rôle in the physiology of the plant. The

cellulose cell-wall of the plant cell is permeable alike to
solvents and solutes, and the protoplasmic lining of the cell
behaves as a single continuous semi-permeable layer round
the vacuole. The result of this is that given, as is always the
case, a cell-sap containing salts and sugars in solution, a cell
placed in water will absorb water without losing cell contents.
It will continue to absorb water and to increase in volume,
pressing against the cellulose wall and distending it as it
does so, until a stage is reached at which the inward pressure
of the cell-wall is equal to the absorbing pressure and no

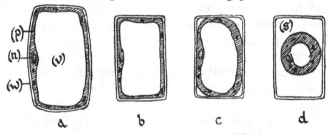

Fig. 3. Diagram to shew a parenchyma cell containing various amounts of water
in the vacuole (*v*). *b*, the cell in equilibrium in a solution of the same osmotic
pressure as the cell-sap in the vacuole. The protoplasm lines the cell-wall
uniformly and the cellulose walls are not distended. *a*, the same cell when
made turgid by placing in water; the cell vacuole has increased in size and
the walls are distended, only their pressure inwards prevents further water
uptake by the vacuole. *c*, the same cell after placing in a solution of greater
osmotic pressure than the cell-sap. The cell is partly plasmolysed, the protoplasm being withdrawn from the cell-walls at the corners of the cell. *d*, the
same cell further plasmolysed; the plasmolysing solution (*s*) passes readily
through the cellulose cell-walls (*w*). (*p*), protoplasm; (*n*), nucleus.

further absorption occurs. If by some means, such as the
conversion of starch into sugar, the osmotic pressure of the
cell should be increased, more water would be taken in and
the cell-wall be further distended until a new equilibrium
would be attained. If a similar cell were placed in a solution
containing a higher concentration of substances in solution
than that in the cell-sap, this solution would readily penetrate the cellulose wall and come in contact with the

protoplasm. Water would then be withdrawn from the vacuole through the semi-permeable protoplasmic membranes. This process, in diminishing the volume of the vacuole, would cause contraction of the protoplasm from the cellulose walls, and the contraction would cease only when the osmotic pressure of the contracted vacuole had been so concentrated by removal of water as to equal that of the external solution. The protoplasm, now far from being inflated against the cell-walls, would be contracted into a mass in the middle of the cell (Fig. 3). The cell would be said to be *plasmolysed*, and the condition so reached is termed *plasmolysis*. If such a plasmolysed cell is put back into water, it takes up water again, the protoplasm presses against the cell-walls which are again pushed outwards, the cell being apparently none the worse after temporary plasmolysis. It was the observation of these phenomena of plasmolysis by the botanist Pfeffer that first directed the attention of physicists and physical chemists to the existence of osmotic phenomena.

It is clear that, having made up a whole range of solutions of different substances, it is possible to put similar plant cells into a series of varying concentrations of each substance. The solution in which the cell is just not plasmolysed is then said to be *isotonic* with the cell-sap, i.e. the osmotic pressure of the solution = the osmotic pressure of the cell-sap. By finding the isotonic equivalents of various chemical substances it is possible to compare the relative effectiveness of various substances in developing osmotic pressures and this was, in fact, the earliest method of investigating osmotic pressures. It is clear that by finding a solution isotonic with it we have a means of knowing the concentration of the solution in the cell vacuole, but this assumes, as up to the present we have done, that the semi-permeability is complete. Actually this is not so. The solute molecules will, in time, diffuse through the protoplasmic membrane and a cell left in strong salt solution will at first plasmolyse and then recover, since the salt will gradually penetrate until the pressures inside and outside the cell are the same. On consideration, it will be realised that such must be the case or otherwise no sugars or salts or any other substance essential for respiration or for plant growth could enter the living cell.

Turgor of cells and of tissues. If a living cell had a wall which was completely extensible, it would expand indefinitely when placed in water, but as the cellulose wall is normally strong and fairly elastic, as the cell is distended the wall exerts an increasing pressure inwards upon the cell vacuole. At length this wall pressure exactly balances the tendency for water inflow produced by the osmotic pressure of the vacuolar solution. In this state the cell is said to be *turgid*; it will be unable to absorb more water and will have a more nearly spherical shape than it had when first placed in water. If the cell is distorted by any external force its volume must be lessened and water forced out of it. The forcing out of water will have to be done against the osmotic pressures tending to bring water into the cell when below the turgor point. Thus definite forces will oppose any tendency to distort the swollen turgid cell, which will therefore have considerable rigidity. The significance of this is less evident in a single cell than in a tissue or plant organ composed of such cells. Let us consider a mass of thin-walled cells with sugar solutions in their vacuoles to be loosely packed into a long cylindrical calico tube. This tube will have in itself little rigidity, and as the cells are loosely packed the whole cylinder of cells will be flaccid and will be easily bent and deformed. On placing such a structure into water all the cells will absorb water until, not the simple pressure of each cell-wall, but the pressure of each cell on its neighbour and the inward pressure of the calico wall on all the cells will prevent further intake of water. In this state the structure will be quite rigid. Any attempt to bend it will mean forcing water out of the cells against their osmotic pressures, and the cylinder, if held at the base, will probably stand upright.

The behaviour of this model illustrates the commonest method by which rigidity is achieved in the stems of small herbaceous plants. The central tissue of living cells is enclosed in a peripheral layer of cells, very resistant to extension though of little rigidity and, so long as water is available and osmotic substances are present in the living cells, the stem will stay rigid and upright. That the tendency of the internal cells to take up water is being opposed by the limiting layers of the stem can be readily shewn by slitting longitudinally into four the cut end of a turgid stem, e.g. daffodil or dandelion. If water is present for them to absorb, the inner cells will now continue to take in water, and will increase in size, so inducing a considerable curvature in the slit segments of the stem, the outer parts of which cannot undergo similar extension (Fig. 4). The curling of the slit portions of the dandelion or daffodil stalk shews the existence of *tissue-tensions* in organs which are kept rigid by turgidity, and such tensions probably exist at various times in all kinds of plant tissues, especially where living cells and dead cells of different extensibilities lie side by side.

If the cells lose water faster than they can get it, as in the case of a seedling exposed to full sunlight or dry wind whilst the root system is in dry soil, then the turgidity will be rapidly diminished and the plant will *wilt*, its leaves drooping and the stem falling over. If this wilting is not too severe or too long maintained, turgidity may be restored when less severe conditions return. This can be seen in the alternate wilting of sunflower plants on hot summer days and their recovery during the cooler nights.

Water could be taken from the cells in another way than by evaporation from the wet cell-surface; that is by placing the whole plant organ in a solution of high osmotic strength, and inducing a condition of plasmolysis in the whole tissue.

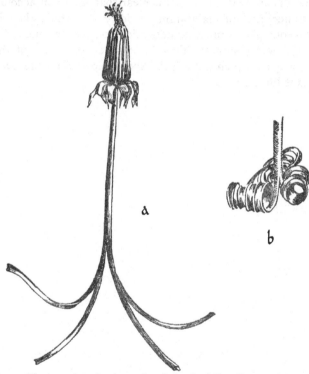

Fig. 4. Tissue-tensions in the stalk of the dandelion flower. *a* shews the curvature assumed by the segments when the stalk is split into four from the base. This indicates that when the stem was whole, the skin was in a state of tension relative to the inner tissues (or the inner cells were in a state of compression). The cutting has permitted the inner cells to expand. *b* shews the effect of putting the same split stalk into water. By the uptake of water the inner cells have expanded so much, whilst the skin has remained relatively unchanged, that the curvature of the split pieces has been very greatly accentuated.

Any cut portion of a turgid stem or root can be rendered flaccid in this way by placing it in a strong sugar or calcium chloride solution. By careful measurement on a uniform tissue such as the pith of a large stem, a distinct diminution in length can be detected. It is clear that solely on account of the necessity of maintaining rigidity, the whole class of herbaceous plants must be dependent on the presence of a plentiful water supply. This is, however, only one of the many ways in which water is indispensable to plant life, even to plant life on land.

CHAPTER III

ORGANIC SUBSTANCES AND THEIR CHEMICAL CHARACTERS

For many centuries, biology has been embarrassed by the view that the chemical activity of living organisms is not of the same type as that in non-living matter. However, during the past century, there has been a gradual change from that position.

The distinction between organic and inorganic chemistry was originally based upon the supposition that organic substances could not be synthesised in the laboratory from their constituent elements. A so-called "vital force" was held to be necessary for their formation. However in 1828, Wöhler obtained urea from ammonium cyanate—an inorganic substance; in 1845 Kolbe synthesised trichloracetic acid and in 1850 Berthelot synthesised alcohol and formic acid. These might justifiably be disputed as mere waste products but Emil Fischer has synthesised innumerable sugars and many of the simpler proteins and it is now even possible to synthesise substances such as thyroxine, which is the active

Adrenaline   Thyroxine

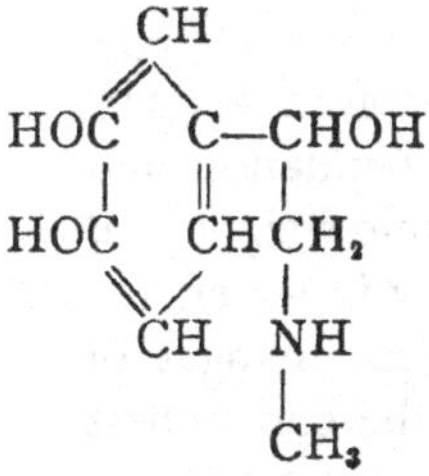

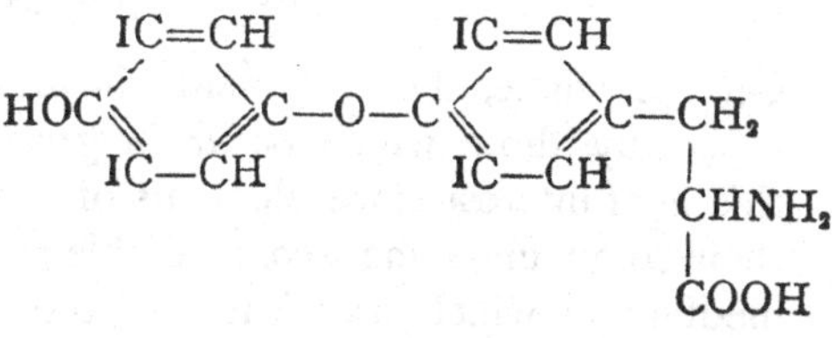

CH
HOC   C—CHOH
HOC   CH CH₂
    CH   NH
        CH₃

IC=CH   IC=CH
HOC   C—O—C   C—CH₂
IC—CH   IC—CH
                    CHNH₂
                    COOH

secretion of the thyroid gland, and which greatly controls the development of the animal organism. Adrenaline, which is also a compound of the utmost importance in the animal body, since the secretion of it controls the blood pressure, has also been synthesised. Organic chemistry has now become the chemistry of the carbon compounds. This truthfully suggests the great preponderance of carbon compounds in the living organism.

A typical analysis of the dried residue of a green plant gives approximately:

Carbon	...	...	45%
Oxygen	...	...	42%
Hydrogen	...	...	6·5%
Nitrogen	...	...	1·5%
Other mineral constituents			5%

There is in a living plant beside this over 90% of water, which indeed occurs in all parts of the plant body. The following table shews the atomic weights of the elements which occur constantly in the plant body and which have to be supplied to the plant for normal development to take place.

H	1·01	P	31·04
C	12·00	S	32·07
N	14·01	K	39·10
O	16·00	Ca	40·07
Mg	24·32	Fe	55·84

On the whole, these elements have low atomic weights, especially those which occur in greatest abundance, and this is of interest since elements of low atomic weight and their compounds shew considerable solubility in water, the medium in which practically all the chemical changes of the plant go on. The fact that carbon is the commonest element in plant tissues is a most significant feature and one which we may explain partly in terms of the very exceptional chemical properties possessed by the carbon atom.

The enormous number of organic compounds known—well over 100,000—testifies to the astonishing power which carbon atoms possess of linking together in great numbers and in many configurations. Large molecules are formed with a central core of carbon atoms and this core carries lateral chains of all kinds, which may readily alter without affecting the central nucleus. The quadrivalence of the atom increases the ease with which substances containing carbon can undergo chemical changes. The possibility of large sized molecules, and the possibilities of varying combinations given by the quadrivalency, allow the carbon compounds abundant structural isomerism. The quadrivalence also makes possible stereoisomerism.[1] There are, for instance, 32 possible isomers of hexose, $C_6 H_{12} O_6$, and as carbon compounds go, this is a small and simple molecule so that the potentialities for variation in molecules as large as those of the proteins would seem to be almost endless. The carbon atom is capable of behaving both as an acidic and a basic radical. Thus it is possible to have both carbon tetrachloride (CCl_4) and the carbonic acid salts ($CaCO_3$). The carbon nucleus may carry

---

[1] *Structural isomerism* can be illustrated by the case of dextrose and fructose; both contain the same number of carbon, oxygen and hydrogen atoms, but these are differently arranged, as is shewn by the fact that one is an aldehyde and the other is a ketone sugar.

Stereoisomerism is shewn by all carbon compounds in which a carbon atom has its four valencies attached to different groups. Note the carbon atom marked © in the formula for a hexose below.

$$CH_2OH-CHOH-CHOH-CHOH-\overset{\overset{\textstyle H}{|}}{\underset{\underset{\textstyle OH}{|}}{\textcircled{C}}}-CHO$$

By using a solid model for the carbon atom and the same four groups, it is possible to build *two* models which are not superposable. These would correspond to two compounds of different chemical properties, and as each carbon atom starred in the formula has also the same 'asymmetrical' potentialities, no fewer than sixteen stereoisomers of this hexose are possible.

at the same time basic and acid groups, groups of oxidising potentialities and groups capable of reducing other compounds. We thus have a picture of the carbon atom, with its enormous chemical potentialities, as the basis of the chemistry of the organic world.

It is interesting to compare silica with carbon in this respect. Its high atomic weight makes it of low specific heat—but far more important, its compounds are mostly insoluble. The complexity of silica compounds is to the geologist what the carbon complex is to the biologist.

In considering the composition of organic material it is necessary always to distinguish carefully between an *organic substance* and a chemical *compound*. The former is generally a *mixture* of various chemicals; its composition and properties are variable and its constituents fairly readily separable. The ordinary cell-wall is such a mixture of substances containing many carbohydrates of various kinds, as well as proteins, fats and other substances. Below are given analyses of cotton fibre, which is the purest kind of cellulose cell-wall, and of wood; both shew a complex composition, which further analyses would shew to be widely variable.

Cotton		Wood	
Cellulose	91·00 °/₀	Cellulose	50—60 °/₀
Fats	·35 °/₀	Lignin	30 °/₀
Protoplasm	·53 °/₀	Other carbohydrates	16 °/₀
Mineral salts	·12 °/₀	Minerals etc.	4 °/₀
Water	8·00 °/₀		

The classes of chemical compounds which occur most abundantly in organic substances are *Carbohydrates*, *Fats* and *Proteins*. The first two classes contain carbon, hydrogen and oxygen and the last carbon, hydrogen, oxygen, nitrogen, sulphur, and sometimes phosphorus.

Carbohydrates. These substances were originally supposed, mistakenly, to be compounds of carbon and water, because the hydrogen and oxygen occur in them in the same proportion as in water. Thus glucose, or grape-sugar, the formula for which is written empirically as $C_6H_{12}O_6$ might have been written as $C_6.6H_2O$. On simple grounds of sweetness, solubility, etc. the carbohydrates can be separated into sugars (including the monosaccharides and disaccharides) and non-sugars (the polysaccharides). The schema below shews the chief classes of carbohydrates and gives common examples of them.

I. Sugars.
- monosaccharides
  - pentoses...Arabinose, Xylose. $(C_5H_{10}O_5)$
  - hexoses...Dextrose, Levulose. $(C_6H_{12}O_6)$
- disaccharides — Sucrose, Maltose. $(C_{12}H_{22}O_{11})$

II. Polysaccharides.
- starch. $(C_6H_{10}O_5)_n$
- cellulose. $(C_6H_{10}O_5)_m$
- xylan. $(C_5H_8O_4)_x$
- gums and pectic materials.

The structure of the simple *hexoses* (sugars with a chain of six carbon atoms) is illustrated below by *d*-glucose and fructose, together with a *pentose* (sugar with a chain of five carbon atoms).

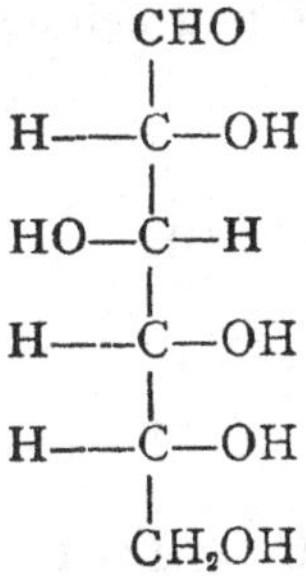

```
    CHO              CH2OH
     |                 |
H——C—OH              C = O              CHO
     |                 |                 |
HO—C—H            H——C—OH           H——C—OH
     |                 |                 |
H——C—OH           HO—C—H            HO—C—H
     |                 |                 |
H——C—OH           HO—C—H            H——C—OH
     |                 |                 |
   CH2OH             CH2OH             CH2OH
 d-glucose.         fructose.         l-Xylose.
```

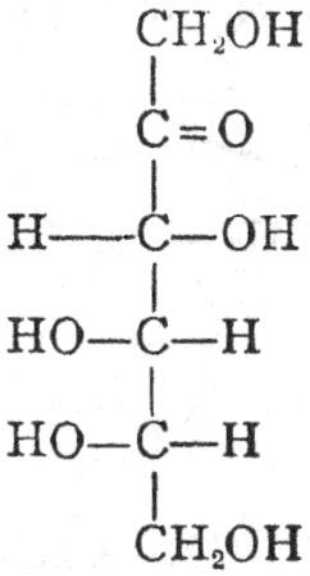

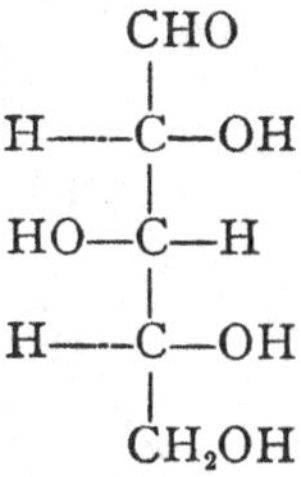

It is almost certain that the simple sugars do not exist in the open-chain form shewn here, but that their atoms are normally arranged to form a ring structure of the type shewn on page 39.

It will be seen that in one case the hexose molecule contains an aldehyde group —CHO, and in the other a ketone group $\overset{|}{C}=O$. It should be remembered that there are fourteen other hexoses different in detailed molecular structure from these, but all having an aldehyde or ketone group in them, and that the *l*-xylose is only one of eight known pentoses all of quite different structure.

The aldehyde and ketone groups are responsible for the reactions in several of the most important tests by means of which the presence of sugars is determined. (Fehling's test, and osazone formation.) These groups are capable of reducing various compounds, themselves becoming oxidised. This is what happens in the case of the test with Fehling's solution where *cupric* oxide is in alkaline solution and is reduced to red *cuprous* oxide by the sugar on heating.

Sucrose, which is a disaccharide, is not a reducing sugar of the above type, probably because it is formed by one molecule of glucose and one molecule of fructose linked up in such a way as to involve both the aldehyde and ketone groups.

CH₂OH—C—CHOH——CHOH—CH—CH₂OH

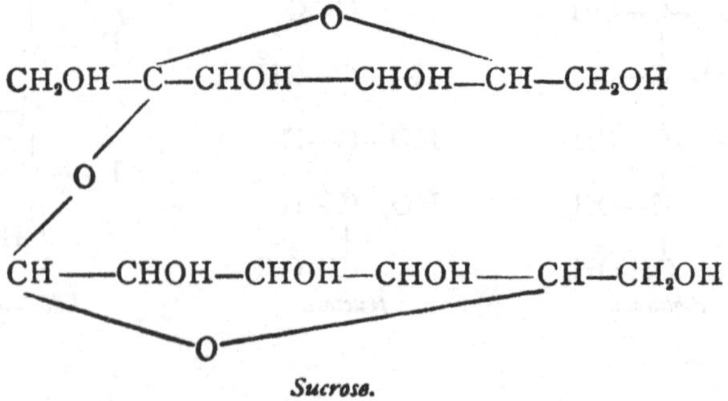

Sucrose.

A very important property of the sugars is their capacity
for condensation in the plant cell to form more complex
molecules Sucrose may be taken as an example of the way
in which sugars may be *condensed* together; *two* hexose
molecules are linked together with the elimination of *one*
molecule of water. In the molecule formed by this condensation $(n - 1)$ molecules of water are eliminated in the
combination of n hexose molecules; the product of such condensation is known as a *di, tri,* or *tetra saccharide* according
to the number of hexose molecules uniting together. It will
be seen that a disaccharide formed from hexoses will have
the empiric formula $C_{12}H_{22}O_{10}$, a trisaccharide $C_{18}H_{32}O_{16}$,
and a tetrasaccharide $C_{24}H_{42}O_{21}$.

As an example of the products of the condensation of
glucose units we may consider the formulae set out below.

a *glucose*
(*monosaccharide*)

a *maltose*
(*disaccharide*)

a-linked glucose chain (*starch, a polysaccharide*)

Here are shewn a molecule of glucose (monosaccharide),
maltose (disaccharide), and below a fragment of the chain
structure of the molecule of starch (polysaccharide). The
molecule of cellulose consists similarly of a chain of glucose
units, arranged however with a slightly different linkage.

Because of the large size of the molecules of polysaccharides they exhibit colloidal properties. This is illustrated
strikingly by the physical properties of the sticky paste
made by heating starch in water, and by the way in which
cellulose takes up water and swells.

Probably because of its high molecular weight starch is
largely insoluble and so forms an ideal reserve material in
which form it is extremely abundant in the plant kingdom.
If it is heated with dilute acid, the starch is converted back
into the simple hexoses, a process known as hydrolysis.
This capability for condensation and hydrolysis in the carbohydrates is of the greatest importance in the cell metabolism
and the same capability is equally important, as we shall
see later, in the proteins. In the plant, however, the processes of condensation and hydrolysis are brought about
not by dilute acids, but by protoplasm constituents of a
highly complex nature, called enzymes. The nature of these
agents is dealt with later (p. 45). Cellulose has a similar
empiric formula to that of starch ($C_6H_{10}O_5$), and, like starch,
upon hydrolysis it yields glucose. Cellulose is important as
the skeleton of the plant body, the substance of which the
cell-wall is made. Linen and cotton fibres are examples
of nearly pure cellulose. Their considerable mechanical
strength is probably due to the fact that the long molecular
chains of glucose units are arranged in parallel along the
length of the fibres. In starch the type of linkage between
the glucose units does not allow a similar formation, the
chains being either spiral or zigzag. Cellulose is not stained
blue by Iodine, and with Schultze's solution (chlor-zinc-
iodine) it takes a purple colour. *Lignin*, which is also an
important cell-wall material, prevalent in wood, contains
condensation products of the pentoses and also aromatic
substances, either of which may be responsible for the staining reactions by means of which this compound is identified.
Its most characteristic reactions are a yellow colour with

Schultze's solution, a bright magenta with phloroglucinol and strong hydrochloric acid and a strong yellow colour with aniline chloride, and aniline sulphate. Gums and mucilages contain polysaccharides which are the condensation products, not of the hexose sugars but of the pentoses.

Fats. The term "fat" is an unfamiliar one in botanical literature and it is more usual to speak of "oil." They generally occur in seeds as reserve products. Generally speaking "oil" means a fat which is liquid at ordinary temperatures, but this use is rather misleading for "fat" has an exact chemical connotation and many oil-like substances in plants do not conform to it.

The fats are produced by the combination of a fatty acid with an alcohol. The organic acids, of course, owe most of the acid properties to the carboxyl group—COOH and the alcohols to the hydroxyl group—OH.

The best known series of acids concerned in making fats is that including the following members:

Formic	$HCOOH$	
Acetic	CH_3COOH	$C_nH_{2n}O_2$
Propionic	C_2H_5COOH	Empiric formula
Butyric	C_3H_7COOH	
Palmitic	$C_{15}H_{31}COOH$	
Stearic	$C_{17}H_{35}COOH$	

Palmitic and stearic acids are the chief components of palm oil, which is much used in soap-making.

Another series of acids has the empiric formula $C_nH_{2n-2}O_2$, e.g. Oleic acid $C_{18}H_{34}O_2$. This occurs as part of olive oil. In the combination of a fatty acid with an alcohol the COOH group and the OH group come together with the elimination of water, just as in typical ester formation:

$$CH_3CO\overline{|OH \quad H|}O{-}C_2H_5 \longrightarrow CH_3COOC_2H_5$$
acetic acid   *ethyl alcohol*   *ethyl acetate.*

In the living plant a very common alcohol is glycerol.

$$
\begin{array}{l}
\text{CH}_2\text{OH} \\
| \\
\text{CH OH} \\
| \\
\text{CH}_2\text{OH}
\end{array}
$$

glycerol.

It has three hydroxyl groups and each will combine with an acid group of some kind. Thus one molecule of oleic acid could combine with each hydroxyl group to give the compound of the structure shewn below:

$$
\begin{array}{l}
\text{CH}_2\text{O} \!-\!\!-\! \text{OC } \text{C}_{17}\text{H}_{33} \\
| \\
\text{CHO} \!-\!\!-\! \text{OC } \text{C}_{17}\text{H}_{33} \\
| \\
\text{CH}_2\text{O} \!-\!\!-\! \text{OC } \text{C}_{17}\text{H}_{33}
\end{array}
$$

This is *olive oil.*

Similar combinations occur between glycerol and the other organic acids and these are true fats. When the carbon chain of the acid is of great length, as in Melissic acid $C_{29}H_{59}COOH$ then the properties of the fat are so changed that it is solid at ordinary temperatures and forms a wax. The properties of the fats are too numerous to be considered now. Their insolubility in water and their capacity for dissolving certain coloured substances not soluble in water, are made use of as tests. They are tested for, microchemically, by means of osmic acid with which they give a dense black coloration, or by means of Sudan III, a stain which is a good deal cheaper and which stains fats red.

There has been much confusion over the part played by fats in the plant metabolism. The fats investigated have been those in food reserves but the more widely occurring types in every living cell are still scarcely if at all understood. A common occurrence of fats or fatty acids, is in the *cuticle* of plants and in *suberin*, the thickening upon the walls of cork cells. The cuticle of leaves when examined in sections stained in Sudan III will be seen to be perfectly distinct from the cell-walls and to stretch like a skin over the outside of the leaf, put on like a layer of varnish above everything else. It is said to diffuse along the cell-walls and spread out over the leaf-surface and there oxidise.

The fats in a seed are undoubtedly formed from carbohydrates and when the seed germinates it is equally certain that simple sugars are formed again. This is a transformation of which the chemistry is still unknown.

Proteins. The protein molecule is the essential basis of the protoplasm. Protein molecules occur of enormous complexity and of great size, and they vary in degree from plant to plant. They have certain reactions by which they are readily identifiable, such as the Xanthoproteic, Biuret and Millon's tests. Whilst it would be impossible to go over the chemical properties of the proteins in any detail, yet the general structure of them is interesting because of some likeness to the carbohydrates and because they play such an important rôle in all living cells. As in the carbohydrates there is a whole range of complexity in the protein molecule, which is now regarded as being built up of smaller units, the amino-acids. The simplest amino-acid is glycine CH_2 (NH_2) COOH, although they all have the NH_2 (amino) and the COOH (carboxyl) groups. It is through these groupings that they are able to condense together like the sugar molecules, with the elimination of water.

$$\underset{NH_2-CH-COOH}{\overset{X}{|}} \quad \underset{HNH-CH-COOH}{\overset{X_2}{|}} \quad \underset{HNH-CH-COOH}{\overset{X_3}{|}}$$

And this process can go on in the protoplasm to form the most complex molecules. Simple "proteins" have already been artificially synthesised by the condensation of as many as seventeen amino-acids. As in the carbohydrates, the complex condensation molecule is capable of hydrolysis, and when this takes place the big protein molecule is split down again into simple amino-acids.

It is worth while noticing that the proteins can behave both as bases and as acids. The COOH group gives them acid properties and the NH_2 group basic properties. To strong bases they behave as acids, and to strong acids as bases. Thus Alanine

$$(CH_3—CH—COOH)$$
$$|$$
$$NH_2$$

forms two sets of salts

(1) with a base

$$CH_3—CH—COONa$$
$$|$$
$$NH_2$$

(2) with an acid

$$CH_3—CH—COOH$$
$$|$$
$$NH_2HCl$$

This capacity, which it is not possible to consider further now, is of the utmost importance in determining the chemical and physiological rôle played by the plant and animal proteins.

Proteins form colloidal solutions in water, and for this reason cannot diffuse out of the cells through the limiting walls of cellulose or of protoplasm. This indicates that the molecule is a very large one, and in fact the molecular weight of the protein molecule has been estimated in various cases to be between 8000 and 200,000. It is a remarkable fact that nearly all the known amino-acids are present in most proteins, so that the properties of the different proteins only vary with the proportions and arrangement of these amino-acid molecules within them.

The cytoplasm of the plant cell contains a large amount of protein as an essential constituent, and exhibits certain properties of proteins in solution. Such protein colloid solutions are very sensitive to physical changes. Thus the cytoplasmic colloids are readily coagulated by heat, the change being irreversible. The action is similar when a cell is killed by a fixative, such as alcohol, formaldehyde, etc., and it seems probable that the life of the cell is possible only so long as the proteins remain uncoagulated. As the degree of dispersion (the proportions of the disperse and continuous phases present in solution) is greatly affected by the hydrogen-ion concentration and salt content of the cells, so the reactions of these proteins largely affect the capacity of the cell for absorbing or giving out water.

The most important constituents of the nucleus are termed nucleo-proteins, and they are extremely complex substances produced by the association of a protein with nucleic acid. The nucleic acid is itself a complex condensation product containing phosphoric acid, sugar and certain nitrogenous bases.

Enzymes and the fluxes of organic materials. From the foregoing survey of the organic materials most commonly found in plants, some idea will have been formed of the number and complexity of the chemical changes going on in the plant metabolism. Especially is it to be noted that each main class of organic compounds contains both simple readily soluble substances and more complex insoluble substances. The latter would tend naturally to occur as storage products or as the permanent components of the plant structure. In each class, carbohydrate, fat, or protein, there must be a constant flux of hydrolysis and condensation between the two types of compound, determined by the external and internal condition of the plant's development. *In vitro*, the hydrolysis of the complex compounds is achieved by dilute acids, but their synthesis is still a matter of great difficulty. In the plant cell (and equally in the animal cell) both hydrolysis and condensation reactions are brought about by substances known as *enzymes*. Each enzyme affects a single condensation-hydrolysis system, so that there is one enzyme (lipase) for the fat to organic-acid and alcohol flux, another (amylase) for the starch to glucose flux and so on. The enzyme can bring about in the cell at ordinary temperatures either reactions it is not yet possible to perform *in vitro* in the laboratory, or reactions which there require high temperatures (that is, at low temperatures they progress very slowly). The enzymes are regarded as *organic catalysts* or as colloidal catalysts. They are not in any sense *alive*, but may be extracted from plant tissues by various means and purified; they will then perform, in a test-tube, the reactions they performed in a plant cell. Such an enzyme is prepared from a pure culture of a fungus and sold under the name of taka-diastase. A description of the preparation of enzymes from yeast is given on page 103.

The enzymes are called catalysts because they have the power of accelerating chemical reactions. The hydrolysis of ethyl acetate to ethyl alcohol and acetic acid is very greatly hastened by a small amount of hydrochloric acid, which is thus said to *catalyse* the reaction,

$$C_2H_5OOC \cdot CH_3 + H_2O \rightarrow CH_3COOH + C_2H_5OH.$$

This reaction is typical of many enzyme reactions and the following points may be noted:

(1) The catalyst does not come into the products of the reaction nor does it alter the final equilibrium, but only hastens a reaction which in time would reach the same equilibrium even in the absence of the catalyst.

(2) The reaction is reversible, and by altering the concentration of reactants or products, it can be made to move in either direction. Thus by increasing the concentration of

ethyl acetate and water, hydrolysis will go on, and by the converse treatment and removing the acetic acid and alcohol, condensation will proceed.

(3) The reaction is a typical case of hydrolysis and condensation, and the one agent is capable of catalysing either process. It has already been said how prevalent this hydrolysis-condensation type of reaction is in the plant body. Among many others, all the following fluxes are going on in the plant body; these represent the main types, and the enzyme concerned with each flux is given in italics.

The enzymes

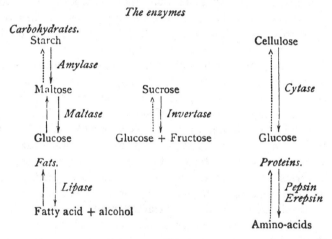

The amylase probably includes a whole collection of enzymes, each capable of one short step in the reaction, but the enzyme lipase can bring about a great variety of reactions, not being limited as in the carbohydrates, and the two enzymes pepsin and erepsin concern the whole range of the protein group.

Although the reactions in this scheme are hydrolysis and condensation reactions, nevertheless there are some other types of reaction and other types of catalyst in the cell.

There are, for instance, oxidases, reductases, etc. Every living cell in the plant contains a whole outfit of these enzymes, and all the different reactions which they promote are going on at once in the cell with varying vigour. Why they do not get inextricably mixed is difficult to understand, since there are no obvious barriers or compartments to be seen within the cell. In some way they must be kept separated by the life of the plant, because when a cell is killed in any way, as when the cells of an apple are killed by cutting or crushing, then autolysis goes on. All the enzyme control, whatever that may be, goes, and a wild confusion of reactions sets in until the whole cell structure is destroyed. The browning of the apple is just one of the most rapid and visible changes setting in.

The enzymes are colloidal, and are readily destroyed by heat, presumably because they coagulate. Their chemical composition is only beginning to be understood in a few instances, but this property of being destroyed by heat (for example, 10 minutes at 80° C.) is a most valuable aid to their identification and separation from inorganic catalysts. The pursuit of the interpretation of physiological processes in terms of physics and chemistry continues steadily, and the last milestone in this progress has been the recent synthesis, *in vitro*, by an extracted plant enzyme-system, of the carbohydrate reserve, starch, which although it occurs so widely in the plant world, had previously never been synthesised save in the plastids of a living cell.

CHAPTER IV

METABOLISM OF THE HIGHER PLANT

It is now necessary to consider how the fluxes of organic compounds, which we have described in Chapter III, are related to the physiology of the higher plant.

The compounds involved, and the materials they are formed from, can be classified into three groups.

(a)	(b)	(c)
Inorganic substances	Simple soluble organic substances	Complex insoluble organic compounds
Carbon dioxide Water Mineral salts Nitrates Phosphates, etc.	Sugars: glucose, sucrose, etc. Amino-acids Fatty acids and alcohols	Starch Cellulose Reserve proteins Protoplasmic protein Nuclear proteins Fats
Raw materials Food of plant	Intermediate products Translocation forms Food of protoplasm Respiration material	New protoplasm New skeleton Reserves

Class (a) consists of those inorganic compounds taken in by the plant and built by it into organic material. These may be called the food of the plant. These in various ways are combined to form the simple organic compounds of class (b), which have small molecules, are soluble, and which are the material out of which the most complex compounds, class (c), are synthesised by enzyme agency. In class (b) are the sugars, amino-acids, fatty acids, and alcohols, and as these are the starting points of the fluxes which end in the most complex of all compounds, namely protoplasm, *these* may be called "*food of the protoplasm*." Thus the distinction is made clear between the food of the protoplasm and the food of the plant as a whole. Not only are the simple organic substances condensed together in this way to form protoplasm but they may form reserve materials of fats, polysaccharides or proteins, which because of their large molecular size are insoluble. Such reserves are laid down in storage organs such as tubers (potato), swollen roots (carrot), and in seeds (pea, bean, etc.).

Further they form the material of the cell-wall, the cellulose and lignin which constitute the skeleton of the plant, though these may in extremities be utilised as reserve food materials also.

The diagram given in Fig. 5 is meant to shew the way in which the processes of change from (*a*) to (*b*) substances, and (*b*) to (*c*) substances, go on in the plant.

At *C* we have the absorption of water and salts in solution, and these pass from the root to the leaves. In the leaf at *A* is illustrated the first synthesis of the class (*a*)→(*b*), namely *photosynthesis*. This process is going on in all the green parts of the plant. Light and carbon dioxide are directly absorbed by the leaf and with water from the roots sugar is formed, as may be roughly expressed by the equation

$$6CO_2 + 6H_2O = C_6H_{12}O_6 + 6O_2.$$

Similarly in the leaf *B* when sugar has been formed, it reacts with the nitrates in solution and in the presence of light, amino-acids are formed. This is also a reaction which can however go on in the dark, although at a much slower rate.

D is a growing point of a stem where cells are dividing, and new leaves, buds, and stems are being formed. To this place in solution, flow the amino-acids and sugars made in the leaves, and salts and water from the roots. There they are synthesised to form new protoplasm—this is the change (*b*) → (*c*).

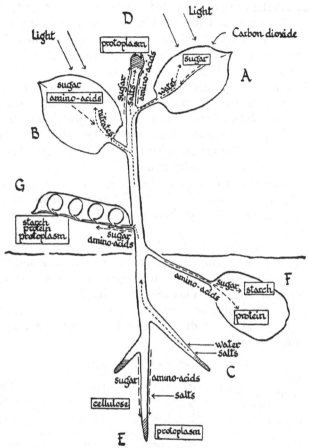

Fig. 5. Schema to shew the chief regions of synthesis of organic substances in the plant. The synthesised product is indicated in each region inside a rectangle and the flow of material to or from the place is shewn by dotted lines, simple inorganic compounds by small dots and simple organic substances by broken lines. *A* and *B* shew photosynthetic processes in the leaf; *C*, intake by the root; *D* and *E*, condensation at growing points; *F*, condensation to form reserves in a tuber; *G*, condensation in the seed forming reserves and new protoplasm.

E. At the apex of the root the same process is going on, and in both root and stem, as the new cells grow, their walls are formed, mainly of cellulose, and as they grow older they increase in thickness by further deposition of cellulose. This also is a condensation or a change $(b) \rightarrow (c)$.

F. The plant in the diagram has been drawn with a tuber. Into any such storage organ flow the simple organic substances in solution, sugar and amino-acids mainly, and these are built up into reserves. The sugar to starch (or sometimes cellulose), and the amino-acids to reserve proteins. This is a condensation, a $(b) \rightarrow (c)$ change.

G. Similar changes go on in the formation of the seeds of a plant, but besides the purely reserve material the protoplasm of the new embryo plant is built up. These are also shewn at *G.*

When conditions in the plant change, then the reserves are hydrolysed again by their respective enzymes and flow away to the regions where they are being consumed and where their concentrations are consequently low.

Thus we are able to view as a whole the chief chemical changes going on in the plant body and we can classify them in this way:

Schema of Plant Metabolism.

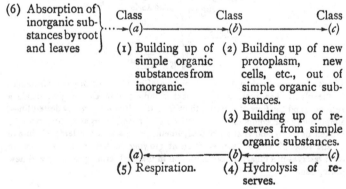

Processes (1), (2) and (3), which are synthetic, we have
dealt with already and of the reverse processes we have
already mentioned the hydrolysis of reserves, (4). To this
we must add the process of respiration. This process is
generally represented as the converse of photosynthesis—it
is the breaking down, by oxidation, of the simple sugars,
and the energy set free is utilised in metabolism, and supplies
the energy for movement and for synthetic processes of change
from (b) to (c). The process is generally represented as

$$C_6H_{12}O_6 + 6O_2 = 6CO_2 + 6H_2O + 677 \text{ kg. calories,}$$

and so fits into our schema as (5).

This respiration occurs in all living plant cells and for
that matter in all animal cells. The dotted line at (6) illustrates the inflow of inorganic material which is the food of the
plant and the raw material of the whole mechanism. This
completes the picture in rough outline of the metabolism of
the plant.

We may now compare it with the similar schema in an
animal.

Schema of Animal Metabolism.

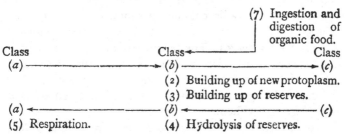

The processes (2), (3), (4) and (5) are similar in essentials
to those of plants, but process (1), as the main source of

the materials of class (*b*), is absent; some synthesis of inorganic substances into organic clearly must occur, but it is small in amount compared with the other processes, and the sugars, amino-acids, etc., of class (*b*) are obtained in an entirely different manner. As is shewn at (7) they come from the complex organic food which the animal eats. The enzymes in the digestive tract carry out the (*c*) → (*b*) processes and so provide the raw material for respiration, the food for new protoplasm, and the source of reserves. The digested foods which yield these simple organic compounds come directly from plants or indirectly from them, since no animal can synthesise from (*a*) to (*b*) by absorbing free light energy in the manner of plants. This schema shews clearly that in plants and animals the food of the protoplasm is the same, but the food of the organism as a whole is different. The schema set out is an incomplete one, but it may be taken as a guide to the more general features of the metabolism of the plant. The term metabolism can best be explained here. It includes all the changes we have spoken of; it is the sum of all the chemical activities of the organism. Those in an upward synthetic direction (*a*) → (*b*) → (*c*) are spoken of as the *anabolic* changes, and the opposite simplifying changes are the *katabolic* changes. In a growing organism the anabolic changes predominate and may be expressed as an increase in new protoplasm, new cell-walls or more food reserves. The term *assimilation* (not carbon-assimilation) is used for the building up of the protoplasm in the stages (*a*) → (*b*) → (*c*), i.e. the incorporation of other substances by the protoplasm and the formation of new protoplasm from them.

Growth. Assimilation must needs result in *growth*. Unless the new protoplasm is continuously destroyed as fast as it is formed it will accumulate and the bulk of the organism will increase. In a simple organism of a certain size, cell division

ensues and two new organisms are formed, each growing to the
parental size. In a complex plant, however, growth goes on
at separate places. At the tips of roots and ends of branches,
as we have indicated, growth is always going on. The
branches both lengthen and thicken. Thus the growth is
localised and *continuous*. In this the higher plant differs from
the higher animal, which latter generally has an earlier
definite growth phase little different for all the tissues, and
no special growing regions. Although the essential characters
of growth are evident, it is often difficult to recognise and
to define. We can scarcely consider temporary inflation as
growth. Growth (that is organic growth) really involves a
permanent increase in the bulk of the organism: not merely
increase in weight, since a piece of gelatine could swell up
by taking in water and it might under constant conditions
lose none: nor simply increase in size—a cell becoming
turgid with water has not necessarily grown.

Perhaps the salient characters of organic growth are
(1) that in growth the organism takes in chemical substances
unlike itself and builds them up irreversibly into its own
substance, (2) that the increase in bulk of the growing
substance means that growth always ends in the reproduction of the individual. In its simplest case this may be seen
in a unicellular plant such as *Chlamydomonas*, in which the
cell grows only to a definite size and then divides.

Differentiation. A further character of growth is that, as
the cells get older and as the organism gets older, *differentiation* takes place. That is to say the various parts of the
organism begin to differ in structure and correspondingly to
play different rôles in the life of the organism. Thus the
simplest organism (e.g. *Chlamydomonas*), is not alike throughout but is differentiated into nucleus, cytoplasm, chloroplast,
cilia, etc., and the nucleus is especially associated with
reproduction, the chloroplast with photosynthesis, the cilia

with locomotion and so on. Thus the differentiation carries with it the principle of *division of labour*, which usually makes the well-differentiated organism more efficient than the relatively undifferentiated ones. In a simple animal like *Amoeba*, or plant like *Chlamydomonas*, differentiation is at a minimum. All the essential processes take place in the one cell, and each cell is "physiologically complete." But in larger "tissue" organisms, different organs and different groups of cells (tissues) are specialised to do different jobs. We have already noted the differentiation of the green plant into regions of photosynthesis, absorption of water and salts, growth, storage and translocation. Here a simile from Mr Henry Ford's writings may help us. Each group of cells is specialised to do a definite job and each workman is specialised to do a definite job. The workmen, Mr Ford says, object strongly to being moved to a new job. They may be even quite incapable of doing the new job, though doing their own superlatively well. Yet Mr Ford's workmen undoubtedly descended from men who not many generations back performed an extraordinary number of varying tasks, year in, year out, in order to keep themselves fed and clothed. They would catch or grow food and prepare it, build houses, make tools, keep animals, make cloth, etc., etc. The cells of higher plants have become like the workmen specialised and differentiated from one another.

As the workmen have specialised so they have increased the efficiency of society, but they have sacrificed their independence and have become dependent on each other. The same is true of plant and animal cells and it is expressed in the phrase "differentiation involves dependence," which is equally as true of all organisms as it is of people.

At this point we cannot begin a discussion of the differentiation of cells, conducting cells, secretory cells, photosynthetic cells, water-absorbing cells, supporting cells and protective cells, but they are dealt with later. For the moment it will suffice that the production of such differentiation is a primary character of growth and is to be seen in consequence, to some degree or other, in all organisms.

A recapitulation of the characteristic activities of the living plant so far dealt with shews the following:

> Feeding and photosynthesis
> Assimilation
> Respiration   } METABOLISM.
> Growth, ending in reproduction
> Differentiation

To these we now have to add a sixth.

Response to stimuli. If the external conditions change, the activities of the protoplasm change also—the protoplasm is said to be sensitive.

In animals the response to stimulus is readily recognisable, for the response is often a rapid movement. In plants the response is less easy to recognise since the resultant movements are slow. Nevertheless they can be readily demonstrated.

A movement in response to stimulus may be made by a translocation of the whole organism, and this is called a *taxis*. If a glass vessel containing rain-water tinged green with *Chlamydomonas* or *Euglena* is placed where light falls upon one side of it, in a few minutes the algae will have swum to that side and will have formed a deep green film there. This movement of a whole organism in response to a light stimulus is called a *phototaxis*.

It is more usual to find the response limited to a movement of a part only of the organism, and such a movement is called a *tropism*. Numerous illustrations could be given of such tropisms, made in response to different stimuli (Figs. 54 and 61). Thus stems of higher plants are usually light sensitive and grow towards the source of illumination. Similarly, leaves turn and place themselves at right angles to the incident light. Both responses are readily observable in window plants and they are both termed *phototropisms*. If stems are equally lighted all round, or if they are grown in darkness, it will be found that they always grow vertically upwards. This is because they are sensitive to gravity, and a stem placed on its side will receive some stimulus which results in its curvature upwards again. This response is called *geotropism*, and it is responsible for the fact that the stems of germinating seeds always come above soil rather than the roots. The roots are similarly sensitive to gravity but they respond by downward growth. The stems are said to be negatively geotropic and the roots positively geotropic. In some cases, plant organs are sensitive to other physical conditions of the environment, such as injury, pressure, water supply, oxygen supply and so forth. Roots, for instance, are positively hydrotropic and so where a tree grows in a dry soil all its roots will grow towards a water supply. It may be a ditch or an open joint in a drain-pipe, and if the latter, as is very common, the whole pipe may become absolutely choked with roots.

All these examples of responses, though slower, are as much responses to stimuli as animal movements and shew the same fundamental properties.

(1) The first of these properties lies in the relation between the stimulus and the response. Generally the re-action is a "trigger reaction," i.e. there is no obvious quantitative relation between the force of the stimulus and the force of the response. In firing a rifle there is little relation between the pressure of the trigger-finger and the muzzle velocity of the bullet. The trigger releases forces, by starting reactions which go on of themselves once started—such as the burning of the powder. And it is exactly similar in

responses of plants and animals. The force which a bean-root exerts in turning downwards through the soil bears no relation to the strength of the gravitational field. The same example may be made to illustrate some of the other characteristics of responses to stimuli.

(2) The character of the response varies with the individual organism, with its condition and type. Thus not all roots are positively geotropic, and although the main root of a plant grows vertically downwards, the laterals may generally be seen to grow at various set angles to the vertical (Fig. 47). Similarly an old root will often shew less response than a young root.

(3) The response varies with the kind of stimulus; thus in the case of the mustard seedling the root grows away from the light and grows towards moisture.

These responses of the plant, and still more the very complex responses of animals to stimuli, may seem at first quite inexplicable in terms of physics and chemistry, but such a mechanism must certainly be involved to some extent and may in many cases be responsible for more than we should think at first. Thus an *Amoeba* on meeting a food particle apparently receives a stimulus and responds by putting out a pseudopodium which ingests the particle. Yet a simple physical explanation might be offered. "The *Amoeba* has a high surface tension and is semi-fluid, and so tends to be round in shape; the effect of the food solution, a simple direct physical effect, is the lowering of the surface tension at the point of contact, and the lowered surface tension causes the extrusion of protoplasm from the *Amoeba* and so the pseudopodium is formed and the particle is enclosed."

A simple non-living model can be made with mercury and potassium bichromate to shew a similar effect. A train of crystals is laid on a dish so that one end of the train touches a big bead of mercury. If strong sulphuric acid is put on the dish, chromic acid is formed which attacks the mercury at the point of contact and lowers the surface tension there. Then the bead pushes out a foot at that point which extends along the chain of particles.

Now there is no doubt that this suggested explanation of the response of *Amoeba* is, in fact, totally inadequate to explain many of its obvious characters, but lowering of surface tension may clearly be involved as part of the mechanism. Such physico-chemical sequences, often vastly complicated, underlie all phenomena of stimulus and response, and biochemistry and physiology are gradually bringing them to light.

Respiration. We have so far considered the process of respiration only in regard to the basic chemical change which is supposed to underlie it, but other more general aspects of it are also of the greatest importance. The chief biological significance of the process rests on the fact that just as photosynthesis is the means whereby energy is chiefly taken in by the plant (and there stored as energy containing organic compounds), so respiration is the essential process whereby such stored energy is set free and becomes available for the complex needs of the plant's metabolism. The necessity for such an energy release is at once apparent. In the case of animals, and plants such as *Chlamydomonas*, energy is used in rapid movement, but in the case of higher plants energy is equally necessary for the slower movements of growth through the soil and up into the air. Whatever may be the nature of the organised control of the cell which prevents autolysis and confusion amongst the enzyme systems of the protoplasm, it is almost certain to involve the using up of energy. A further very large expenditure of energy must come in the synthesis of the complex organic compounds in the plant from simpler materials of smaller potential energy content.

As we have already said, this release of energy is regarded as due almost entirely to an oxidation within the cell of a hexose sugar to carbon dioxide and water. It is possible that under conditions of starvation other substances may be similarly oxidised, but the evidence in favour of the normal substance being glucose is now quite strong. The reaction can be empirically stated as

$$C_6H_{12}O_6 + 6O_2 = 6CO_2 + 6H_2O + 677 \text{ kg. calories.}$$

It clearly implies a number of exchanges between the plant cell and its external environment. Thus there will be an intake of oxygen and an equivalent output of carbon dioxide. This can be demonstrated experimentally by passing a very slow and constant stream of air over a mass of living plant material (kept in darkness if it is green, in order to avoid photosynthesis) and by analysing samples of the gas stream both before and after passing through the closed respiration chamber. It would then be found that whilst the oxygen content of the air would have diminished by one or two per cent. (e.g. to 19·5 $°/_o$) the carbon dioxide content would have correspondingly increased (i.e. to 1·5 $°/_o$). This gas exchange can also be demonstrated by placing germinating seeds, which respire rapidly, in a vessel containing caustic potash and closed by a manometer. The carbon dioxide produced would be absorbed and the continual using up of the limited oxygen supply would cause negative pressures to be registered by the manometer. Many other pieces of apparatus of a similar kind are in use. The oxygen used in respiration is present to the extent of 21 per cent. in ordinary air and it reaches the respiring cells by free gaseous diffusion through the skin and air-spaces of the plant tissues. It finally dissolves in the water which saturates the cellulose walls of the cells, and by hydro-diffusion passes through them and through the protoplasm. Its continual consumption inside the cells maintains low concentrations of the gas inside the cells and so keeps going the diffusion processes. Similarly the production of carbon dioxide in the cell raises the internal concentration of that gas, with the result that it continually escapes to the outside air by the same path and the same diffusion processes as those by which the oxygen enters. In any green organ in the light, the converse gas exchanges to those of respiration are produced by photosynthesis, and as this latter process is usually much the more rapid, the respiration gas exchange will be masked. Nevertheless the respiration process must be regarded as going on all the time in such organs, even though it only becomes apparent in the absence of light.

It is more difficult to shew that sugar is consumed in respiration than to shew the gas relations, but analyses under different conditions reveal a surprisingly consistent parallelism between the sugar concentrations and the respiration rates, and in analyses of starved leaves (i.e. leaves not permitted to photosynthesise and forced to use up their reserves) the amount of sugar lost by them corresponds closely with the quantities of carbon dioxide given out.

Whilst some of the energy set free by the oxidation of sugar is used in the plant metabolism directly, some at least appears as heat energy. The rate of production is, however, so small that unless steps are taken to conserve it, only an undetectable small increase in temperature occurs. If, on the other hand, rapidly respiring plant material such as germinating peas is kept inside a vacuum flask, a rise of many degrees in the temperature will soon be shewn above that in a comparable flask containing dried peas or peas which have been killed by heating.

The demonstration of the production of water in respiration is a much more difficult matter on account of the abundance and variability in amount of this compound in the plant.

The significance of the respiratory process to the plant is an extremely deep one; it is the basis on which the whole of the developmental processes of the plant must rest, and it is worth noting that a high respiratory activity seems to occur in all organs in an active state of growth, whilst a low respiration characterises the dormant condition of seeds, tubers, etc. The respiration of the plant differs from that of the animal not so much in the underlying chemical reactions, as in the fact that every plant cell takes in oxygen directly from air in the plant air-spaces, and gives carbon dioxide directly into them, whilst the animal has a complex system of lungs and a circulatory blood-system whereby the transport of these gases to all the cells and their exchange with the outside air is much facilitated. It is the behaviour of these organs in the animal which gives rise to "breathing" as distinct from respiration. No such organs are present in the plant, nor should the term "breathing" be applied to its respiratory processes.

Nor should it be finally overlooked that although the majority of biologists consider the significance of plant-respiration in the way above indicated, the ascertainable *facts* of that process are limited to the manifestations of the biochemical processes we have mentioned: most if not all of the liberated energy is lost in heat production, and the idea that there is also a utilisation of energy set free in the process remains no more than a hypothesis.

CHAPTER V

THE PLANT CELL

The cell unit. The cell is the structural unit of the organism whether animal or plant, and whether unicellular like *Chlamydomonas* or multicellular like the vast majority of more complex organisms. These cell units tend to differ in the higher plant and higher animal in that the plant cell is limited by a rigid skeletal wall of cellulose or lignin, a structure which persists even when the cell is dead. The animal cell on the other hand consists frequently of naked protoplasm and whilst there is no rigid cell-wall, in some tissues, such as cartilage or bone, an intracellular non-living matter is secreted from the cells. This is not layered like a wall but forms a more or less homogeneous matrix. This distinction between the plant cell and the animal cell emphasises the fact that the cell unit consists essentially of a mass of living protoplasm divisible into cytoplasm and nucleus, whilst the cell-wall structures, so evident in plant tissue that their resemblance to the cells of a honeycomb was the actual origin of the term "cell," are not the essential, but the incidental constituents of the living cell. The term "cell" has now come to have a very wide application and is used of the living protoplasmic units both with and without cell-walls, and is also given to the dead cell-wall skeletons. Whilst it is generally true that all organisms are composed of cells, some, such as the fungi (e.g. *Mucor*), are *coenocytes*. The plant body is composed of a long branched tube with walls made of a special kind of cellulose, and with no cross-partitions. The protoplasm lining this tube contains numerous nuclei spaced apart, and these multiply at the apices of the

tube where extension and the formation of new cell-wall
and new protoplasm is proceeding. Such a construction is
unusual however and the majority of plants consist of discrete cell units each with a cellulose wall. Every such plant
has begun its development as a single cell, for cells do not
arise *de novo*, but originate only by the division of pre-
existing cells. By the continued division, growth and differentiation of a single initial cell, cells in vast numbers are
produced and give rise to the various tissues and members
of the adult plant body. The cells which thus occur are
said to constitute a *tissue* when a number of them have some
common character, such as structure (e.g. fibrous tissue),
function (e.g. conducting tissue, supporting tissue), or position
(e.g. cortical tissue). Though all the cells possess individuality, such that each equally respires and deposits its own
cell-wall and remains recognisably distinct from its neighbours, yet all the cells of the organism are inter-related and
organised into the single unit of the organism. They are not
merely a crowd of individuals inside a common skin, but an
organism which has an existence as a unit, reproducing as
a unit, responding to stimuli as a unit, and shewing differentiation and division of labour among its constituent
elements.

The origin of new cells by division from older ones goes
on more or less simultaneously all over the body of the
animal so that all the organs of a baby animal appear to
develop at much the same rate, but in a plant there are
localised regions (meristems) in which the dividing activities
of the cells persist after they have been lost by the main bulk
of the plant. Such meristems are mainly found at the apices
of roots and stems which therefore continually extend in
length and there all the cells are "young" because they have
just been formed. They are spoken of as meristematic or
embryonic cells. The presence of growing *regions* in the

space organisation of a plant seems to go with the absence of any special *time*-phase of development, which proceeds all through the life of the organism. In the animal which has no growing regions there is a much more definite growth *period* in the development of the organism.

In a microscopic section through the meristematic region of a plant stem or root the young cells have a very characteristic appearance (Fig. 6). The cell-wall is extremely thin, the cell is more or less cubical in shape, the cell cavity is full of protoplasm containing no vacuoles and the nucleus is very large and conspicuous, with a diameter up to two-thirds that of the cell. The granular cytoplasm contains small deeply staining bodies, the plastids, which may develop in the older cell into chloroplasts or leucoplasts.

The nucleus. The nucleus of the living plant cell is a spherical body with a *nuclear membrane* containing a network of *chromatin* (so called because of the readiness with which it takes up stains). Embedded in it can be seen a round, deeply staining body which is called the *nucleolus*, and which is held to be of the nature of a reserve of chromatin. The nucleus in this condition is said to control and direct most of the metabolic activities of the cell, a view put forward on general grounds but supported also by a certain amount of direct evidence. Thus when certain uninucleate organisms or cells are cut into two parts, one nucleate and the other enucleate, only the part containing the nucleus will persist and regenerate. The enucleate part may live for some time but will not grow or divide, and cell-wall formation will not take place in it. Circumstantial evidence to the same effect is given by the movement of the nucleus to those parts of a cell in which metabolic activity is greatest, for example to a region of cell-wall extension or of fungal invasion.

Nuclear division. The cell undergoing division in a meristem shews a most marked nuclear behaviour, which

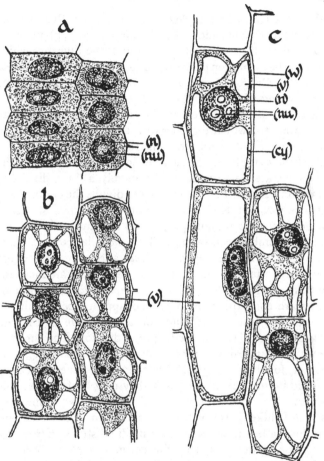

Fig. 6. Extension and vacuolation of cells. *a*, cells from the meristematic region of a root (i.e. the region of cell-division) shewing small thin-walled cells with no vacuoles. *b*, cells from 2 mms. behind the meristem, droplets of cell-sap forming and coalescing to form vacuoles; the cells are larger and the walls thicker. *c*, cells from 7 or 8 mms. behind the meristem, shewing more advanced vacuolation. (*n*), nucleus; (*nu*), nucleolus; (*cy*), cytoplasm; (*w*), cell-wall; (*v*), vacuole. (After Sachs.)

although extremely complex, is carried out with the most extraordinary constancy. In all animals and plants it is substantially the same, differing only in the divisions which precede the formation of the reproductive cells. This process of nuclear division is called *karyokinesis* or *mitosis* (Fig. 7). The staining of a living plant cell usually shews the nuclear structure described above, namely a network of fine material, the chromatin, which together with the nucleolus has taken up the stain more readily than the rest of the nucleus, or the rest of the cytoplasm. Such a nucleus is said to be in the "resting state," a term implying merely a "non-dividing" condition. As the nucleus is about to divide, the chromatinic network resolves itself into a number of strands still conspicuous by their staining properties; and the nucleolus and nuclear membrane disappear. The short lengths of chromatinic material are the *chromosomes*, and even at this stage they may often be seen to have developed an incipient splitting along their length into two *chromatids*. The chromosomes are constant in number and appearance for all the cells of any species of plant or animal save in the cell divisions immediately preceding reproduction. The chromosomes now appear to enter a phase of attraction for one another, for they shorten and thicken, and come to occupy the middle of the cell in which they now lie free. As this happens, the cytoplasm of the cell develops two systems of fine threadlike fibrils which do not stain and which are therefore called the achromatic figure (in distinction from the chromatic figure formed by the chromosomes). The fibrils radiate from two centres (poles), one in each half of the cell, and as mitosis proceeds they extend, especially inwards, until the two systems meet and form the shape of a spindle. The widest part of the spindle lies across the middle of the cell and on this plane the chromosomes become arranged. Some of the spindle fibres pass right through from one

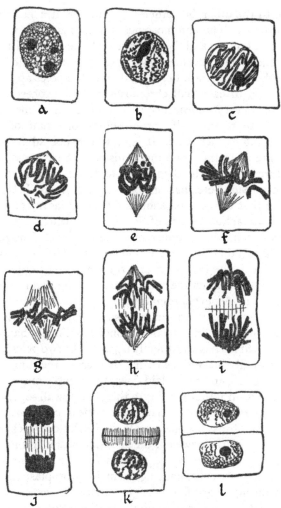

Fig. 7. Mitosis in the meristematic cells of the root tip of the onion. Stages
a to *d* shew the formation of the chromosomes and the disappearance of the
nuclear membrane, *e* shews the spindle-fibres formed, in *f* longitudinal splits
appear in the chromosomes (first visible in *c*), in *g, h* and *i* the half chromosomes are separating and in *j, k* and *l* they are reconstituting two daughter
nuclei. Cell-wall formation is first apparent in *i*. (After Buehner.)

spindle to the other, but others appear to be attached to the chromosomes. The chromosomes are at this stage usually bent into a U-form and the fibrils seem to be attached to the bend. Each chromosome now shews a definite split along its length and separate fibrils leading to the poles appear to be attached to each half chromosome. The process which follows has every appearance of a contraction of the fibrils dragging apart the half chromosomes, one half of each chromosome to each pole. It is, however, not advisable to be definite as to the exact mechanism of the process since the nature of the fibrils themselves, as "lines of flow," "cyto-plasmic threads," etc., has not been agreed upon. It is most probable that the phenomenon is due to replacement of the phase of attraction by one of repulsion, in which the chro-matids of each chromosome tend to move apart. The result in any case is that the comparable sets of half-chromosomes move towards the two poles and there each set merges into a close mass in which the individuality of the chromosomes is lost again. A new chromatinic network, a nuclear mem-brane and nucleolus appear, and so at each pole arises a "daughter nucleus" in a "resting" condition. As this happens, the spindle fibres passing directly between the two poles begin to shew small thickenings where they cross the equatorial plate of the cell, and these, laid down in a double manner, in-crease in size until they fuse into a thin double wall separating the cell into two halves each containing one daughter nucleus. Most of the fibrils disappear as further cellulose thickening is deposited by the new cells on each side of the first formed "middle lamella," but some threads remain in continuity between the two cells, keeping them in protoplasmic con-nection through very fine pores or pits in the walls. These are the protoplasmic connections mentioned on p. 9.

Nuclear division in animals differs little from the process described above. One of the most frequent differences is that the animal nuclear division is preceded by the division of a subsidiary body, the blepharoplast, into two centrosomes which move to the two poles and initiate spindle-fibre formation, and which retract into a blepharoplast when division has been completed. Such bodies are generally lacking in the plant cell. Another usual difference between the cell division of the two types of organism is that the new cross wall is formed in the animal cell by the ingrowth of the walls of the dividing cell as a "cleavage furrow" which, like a closing iris diaphragm, eventually cuts apart the two newly formed nuclei into separate cells. The distinctions between the cell divisions of animals and plants are nevertheless not great, nor constant (Figs. 8, 22 and 30).

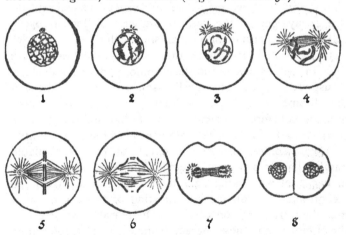

Fig. 8. Nuclear division in an animal cell. Diagram for comparison with plant cells. The chromosome behaviour is closely similar in the two cases, but in the animal cell a blepharoplast division precedes spindle formation and the daughter cells become separated by the development of a cleavage furrow.

The two types agree remarkably in essentials, and in these the most striking thing is the equality of the division of every chromosome by longitudinal splitting. A half, an exact half,

goes to each daughter nucleus. The fairness is obviously of
the greatest importance if, as we have reason to think, the
nucleus really does dominate the cell metabolism. Its
significance is even greater if we regard the nucleus as the
means by which hereditary transmission occurs and if we
imagine every chromosome to have arranged along its length
all the units controlling the hereditary properties of the
organism. And this is actually what is supposed to be the
case. It is common knowledge that by taking a few cells of
a plant, from root, or leaf, or stem, quite ordinary vegetative
cells, a new complete plant can be made to grow. This
means that every vegetative cell has in it all the potentialities of a whole new plant, and indeed of a whole race of
plants, and since we may take any vegetative cell and do
this, always getting the same result, we may look in every
cell for some permanent mechanism which we may consider
as the possible bearer of all the hereditary qualities. Here
then, in the nucleus and in the chromosomes of the nucleus,
is just such a permanent mechanism. At every cell division
the chromosomes reappear. Each new cell has the same
number as before. In every plant of that species in every
smallest part of root or leaf, every cell has the same number
of chromosomes. Every single cell of this incalculable
number has its proper complement like every other cell, and
in each cell the same definite shapes and sizes can be
recognised. The longitudinal splitting of the chromosome
must be a true division of all the hereditary characters of
the organism and these characters must be arranged in units
longitudinally down each chromosome, the unit bearing
each character being neatly split into two at the time of
mitosis.

In all but the specialised reproductive cells of a plant or
animal the nucleus contains one chromosome descendant of
each and every chromosome present in the first cell of the

series, which was formed by the fertilisation of a single egg cell by a male nucleus.

Although the chromosome theory of the transmission of hereditary characters which we have outlined has so much to recommend it, and is generally accepted, yet one major argument can be held against it. This is that *chromosome persistence* in between one nuclear division and the next is hard to verify, since the chromosomes lose their staining properties during the resting phase of the nucleus. Until a year or two ago all the complex processes of mitosis had been deduced from stained sections of tissues in which the cells had been killed in all the various stages of division. From them the process had been reconstructed. Within the last few years Dr Strangeways, using animal cells in culture solutions, has been able to follow under the microscope all the changes going on in a living cell from the beginning to the end of cell division. His work has very thoroughly substantiated the older deductions as to chromosome behaviour in the nuclear division itself. In between one division and the next, however, the chromosomes do not stain. Though fourteen chromosomes may emerge from the resting nucleus at one nuclear division as fourteen went in after the preceding division, and though the shapes are the same, there is no actual proof of chromosome individuality during the resting state, though all the circumstantial evidence points towards it.

Apical growth. In the region of *cell division*, the cells are constantly multiplying and since these do not get progressively smaller as the plant apex grows older, but remain of much the same size, it follows that there is a great formation of new protoplasm going on in this region. Such indeed we have seen to be the case, in that soluble organic materials from leaves or storage organs constantly move to the growing point and are there synthesised into new protoplasm and new cell-wall. The constant formation of young cells causes the root or stem apex to move slowly forward. As this happens, each cell becomes further and further behind the growing point and its distance from the dividing region will be roughly a measure of its age. With increasing age, the cells go through a cycle of varying metabolic activity. The youngest cells in the dividing region itself are in a phase of protein condensation, but slightly behind this zone it becomes evident, on careful examination, that the tendency of the cell metabolism has changed, for small granules of starch begin to be visible in the protoplasm. This phase does not last long in the developmental history of the cell and the starch zone in the root tip is short. Immediately behind it, the cells are seen to undergo an enormous and sudden alteration in size. Drops of liquid sap form in the cytoplasm, and grow and coalesce, forming a vacuole inside the cell. The nucleus remains suspended in this vacuole by bridles of protoplasm. As the increase in the vacuoles goes on, the cells become greatly distended. This is due to the fact that the cells achieve a considerable turgor. This phase coincides with the disappearance of some or all of the starch granules from the cells, and it is most probably the hydrolysis of this starch to sugar, by the changed metabolism of the cell as it becomes older, that is responsible for the increased turgor (Fig. 48). The sudden appearance of large amounts of sugar in the cell is responsible for high osmotic pressures which "draw in" water through the semi-permeable protoplasmic membranes and so form vacuoles which continually increase in size and finally coalesce. The pressure of the absorbed water distends the cell practically to breaking point and all the time the protoplasm continues to lay down fresh cell-wall material in, and upon, the stretched wall. Thus the constantly distended cell-wall is always growing rapidly in size. This enlargement is not equal in all three dimensions of the cell, but is greatest longitudinally. In the adult and elongate cell, the protoplasmic bridles have collapsed and the protoplasm lines the cellulose walls in a thin layer in which the nucleus may be seen at one side of the cell (Fig. 6).

Although the osmotic mechanism is responsible for the pressures which extend the cells, it is now known that the onset of extension is determined by a great increase in the plasticity of the young cell-walls, and that this in turn is due to growth-controlling substances moving backwards in small amounts from the apices (see p. 244).

This *phase of elongation* is probably the most marked period in the growth of the root or stem so far as outward appearance goes. A well-known experiment serves to demonstrate this point. The main root of a bean seedling is washed clean and wiped dry and is then marked off by lines of Indian ink, into equal spaces of about one millimetre, beginning at the apex of the root. The root is moistened and the seedling is left to continue growth under suitable conditions. When examined again, after 24 or 48 hours, it will be found that the first one or two spaces at the root apex have increased little (the root cap and meristematic and starch-forming zones), the next two or three spaces will have increased to many times their original length, and the spaces further and further back will shew progressively smaller and smaller extensions (Figs. 54 and 58). If a longitudinal section is cut through such a root, and is examined under the microscope it will be found to verify the fact that the region of elongation as seen externally coincides exactly with the region of the internal phenomena of cell-vacuolation and extension. The table given below illustrates the results of a marking experiment similar to that described, the material used being a broad-bean root marked off into ten consecutive spaces of one millimetre, and a runner-bean stem marked off into twelve spaces of three and a half millimetres each:

Vicia root. Increments on each of ten 1 mm. spaces

1·5, 5·8, 8·2, 3·5, 1·6, 1·3, 0·5, 0·3, 0·2, 0·1 mms.

Phaseolus stem. Increments on each of twelve 3·5 mm. spaces

2, 2·5, 4·5, 6·5, 5·5, 3·0, 1·8, 1·0, 1·0, 0·5, 0·5, 0·5 mms. (apex of plant).

Thus it is clear that only the apical centimetre or so of a root is actively elongating and the region a few millimetres from the root tip is the region of greatest elongation. In the stem the elongating region is more spread out but the region of greatest elongation here also lies close behind the meristem. The forces developed by the elongation process are very considerable and may amount in a growing root to a hundred pounds pressure to the square inch. Such large forces must clearly be present where roots penetrate stiff clay soils or where their growth lifts up paving stones or asphalt pavements, occurrences quite commonly met with.

The cell-wall of the elongate cell continues to thicken and cellulose is deposited in successive layers upon the original wall or "middle lamella." This original wall is different in composition from the later deposited wall and usually remains visible in all sections through the walls of cells.

Each cell has its own discrete wall. At certain spots the cellulose thickening may not be laid down and then only the original lamella remains between each cell and the next. Such a space is called a "pit" and these thin places in the walls facilitate the passage of solutions from cell to cell.

In this region of cell deposition, the mature structure of the various cells of stem and root tissues is determined. Some are elongate and open conducting cells, and some are thick walled and tapered mechanical fibres, and so forth. Thus this third zone is said to be the region of *differentiation* or of *maturation*. At a later stage we shall have to consider the different types of cells which result from these activities of cell division, extension and differentiation.

CHAPTER VI

PHOTOSYNTHESIS

We have already said that the plant synthesises all its complex organic substances from carbon dioxide, water and inorganic salts, and that the initial metabolic process from which all the others begin is called photosynthesis. This is the formation of a carbohydrate such as sugar or starch from carbon dioxide and water. It can only take place in the presence of light and through the agency of the green pigment, called chlorophyll. In the process carbon dioxide is absorbed from the ·03 to ·04 per cent. present in the air, oxygen is given out, and light is absorbed.

Probably the necessity for light absorption and the absorption of carbon dioxide in such dilutions, is connected with the fact that the assimilatory organs of plants are broad, flat, widely spread structures which may sometimes be flattened stems, but which are generally leaves. The majority of the cells in the green leaf are similar in appearance and are capable of photosynthesis. The branching vein system of the leaf, which is concerned with the conduction of water and food materials and with the support of the leaf, and the single superficial layer of cells on each surface of the leaf, the epidermis, are both non-photosynthetic. The rest of the cells of the leaf, the mesophyll, is entirely photosynthetic tissue, although usually divided into two regions of which that next the upper surface, the palisade layer, is the more active and the lower layer, the spongy mesophyll, the less active (Figs. 12, 63 and 65). The mesophyll cells, especially the palisade cells, contain densely packed chloroplasts. These are protoplasmic structures of a colloid nature which contain a great deal of protein and large amounts of fats, both of which are regarded by some people as of the nature of food reserves. The chloroplast body is usually lens-shaped and a shell of the pigment chlorophyll covers the body (stroma) of the plastid. These plastids are not formed spontaneously from the cell proto-plasm, but arise during cell division by a corresponding division of the small plastids present in the cytoplasm. Such plastids grow and develop, sometimes into chloroplasts and sometimes into colourless plastids, the "leucoplasts" of colourless plant organs. The chloroplasts are free to move in the cell, and may move with the streaming protoplasm as in the cells of *Elodea* (Canadian Pondweed). They also shew some independent power of movement and in the palisade cells, in dull light, move towards the upper and lower cell-walls, crowding densely over them, whilst in bright light they mainly occupy the side (vertical) walls of the cells. The mechanism of such movement is not in the least understood. The chlorophyll which covers the chloroplast surface is present in a colloidal condition and though readily extractable from the leaves in ethyl alcohol, acetone, chloro-form, ether, etc., the extracted pigment is not quite similar to that on the chloroplast and it will no longer photosyn-thesise. The easiest method of extraction of pigment from a leaf involves plunging the leaf into boiling water till flaccid. This is supposed to send the chlorophyll into true solution in the fats of the chloroplast, and they can then be easily extracted by the fat solvents mentioned above.

Although it is a mixture of pigments, analysis of the chlorophyll from plants belonging to all parts of the plant kingdom shews that an astonishing uniformity of composi-tion exists. In all green plants the four pigments mentioned below are present, and in plants such as the brown and red seaweeds the same pigments are present though masked by large amounts of the other brown or red substances.

*Typical analysis of the Chlorophyll Pigments present
in the Green Leaf.*

		per cent. of dry weight of the leaf
Chlorophyll α	$C_{55}H_{72}O_5N_4Mg$	
microcrystalline, green-blue		0·63
Chlorophyll β	$C_{55}H_{70}O_6N_4Mg$	
microcrystalline, pure green		0·24
Carotin	$C_{40}H_{56}$	
crystalline, orange-red		0·05
Xanthophyll	$C_{40}H_{56}O_2$	
crystalline, yellow		0·09

The carotin is found separately as the ordinary colouring matter in carrots, in flower petals and in the eyespot of such algae as *Chlamydomonas*, whilst xanthophyll occurs in autumn-yellow leaves. The two green pigments do not however occur separately. In view of the rapid photochemical reaction which they promote it is interesting to note the presence of magnesium in both chlorophyll α and chlorophyll β. Iron is not present in the pure pigments although necessary for their development in the plant.

Absorption of Energy. The heat of combustion of one gram of glucose to carbon dioxide and water is $3·76 \times 10^3$ calories and as photosynthesis must carry out the converse process, for each gram of sugar formed in the leaf $3·76$ kilocalories of energy will have to be absorbed. If another carbohydrate than sugar is formed, the value will be a little different. This energy is taken in by the chlorophyll although light of varying wave-lengths is not equally well absorbed. The green colour of chlorophyll seen by transmitted light indicates that from white light falling upon it the red rays are absorbed and the green allowed to pass through. A more exact idea of the absorption can be obtained by placing a chlorophyll solution between the source of light and the slit

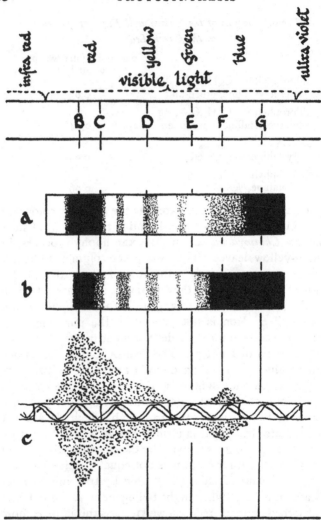

Fig. 9. Diagram to shew the effect of light of different colours upon photosynthesis. At the top of the diagram are shewn the spectrum colours and the lettered vertical lines are the characteristic Fraunhofer lines of the spectrum of

of a spectroscope. In the spectrum, absorption bands will be present, indicating by their density the degree of absorption of light of the wave-length of that part of the spectrum. The absorption spectrum given by a solution of chlorophyll differs little from that given by the living leaf. In each case the following result is obtained—that there is the greatest absorption in the red and blue parts of the spectrum, whilst absorption in the green part is very small (see Fig. 9). It remains to be shewn that the light so absorbed is utilised in photosynthesis and two simple methods of demonstrating this have been employed.

(1) *Engelmann's experiment.* A filament of a green alga, such as *Spirogyra*, capable of carrying on photosynthesis in water, is mounted on a microscope slide in water below a coverslip. In the water is placed a rich culture of a bacterium (*Bacterium termo*) which is sensitive to oxygen and moves towards any source of oxygen. A spectrum is then thrown longitudinally upon the algal thread, which photosynthesises and gives off amounts of oxygen into the water varying at different points with the intensity of the process. The bacteria swarm towards the regions of greatest oxygen production and shew by their accumulation the regions in which photosynthesis is going on most actively. As Fig. 9 shews, the regions of maximal photosynthesis corre-

---

sunlight. *a*, shews the absorption spectrum of a solution of the two green chlorophyll pigments extracted from a leaf; *b*, the absorption spectrum of a living leaf; the two maximal absorption regions, in the red and blue, are seen to approximate in *a* and *b*. *c*, diagram to illustrate Engelmann's experiment. An algal filament illuminated along its length by a spectrum photosynthesises and produces oxygen; the region of greatest oxygen production is seen to be the red and orange corresponding with the region of greatest absorption by the chlorophyll. A secondary maximum occurs in the blue end of the spectrum, the fact that it is smaller than the maximum in the red is due to the much smaller content of energy in the blue end of the solar spectrum. (*a* and *b* after Willstätter and Stoll.)

spond with the regions of maximal absorption by the chlorophyll.

(2) A similar rough demonstration of the same point can be made by keeping a plant in the dark until quite free from starch, and then throwing upon it a strong spectrum. After some hours of photosynthesis the leaf is detached from the plant, and the chlorophyll having been extracted from it, it is stained with iodine. All the leaf will be colourless save those parts of the spectrum which brought about photosynthesis and in these a deposit of starch will appear as a deposit stained blue-black by the iodine. The efficiency of the light absorbed will be shewn by the depth of colour produced. It is interesting to note that by a variation of the method of experimentation, it can be shewn that radiation beyond the wave-length of visible light, namely the infra-red, can be used to some slight extent in photosynthesis. Probably any of the absorbed light can be used in photosynthesis, and it is the amount of the different wave-lengths absorbed which controls the amount of photosynthesis in different parts of the spectrum.

Measurement shews how small a part of the light falling upon a leaf is actually used in photosynthesis.

It has already been indicated that the generalised equation for the photosynthetic reaction can be expressed as

$$6CO_2 + 6H_2O + Energy \longrightarrow C_6H_{12}O_6 + 6O_2.$$

The stages by which this reaction proceeds have, however, been disputed. It is widely believed that formaldehyde is formed as an intermediate product in the reaction, and that it immediately polymerises to form sugar molecules. Thus the reaction should perhaps be written

$$CO_2 + H_2O + E \longrightarrow CHOH$$
$$6(CHOH) \longrightarrow C_6H_{12}O_6$$

Immediate polymerisation of the formaldehyde has to be supposed since this substance has never been identified in the living cell, to which indeed it acts as a strong poison. The first identifiable product of photosynthesis is usually glucose, but in many plants, though by no means all, starch is formed from the hexose as soon as it is produced by the

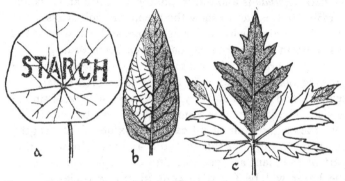

Fig. 10. Starch formation and removal from the leaf. *a*, a Tropaeolum leaf covered with a black paper stencil and exposed to light for two or three days has then been gathered, decolourised and stained with iodine. In the parts of the leaf to which light had access *via* letters cut in the stencil, starch has formed, so that the letters STARCH stand out black on a whitish background. *b*, a privet leaf stained in iodine like leaf *a*, starch is present in all parts of the leaf save a section from which carbon dioxide was excluded by covering with cocoa butter during exposure to the light. *c*, a leaf similarly stained in iodine after remaining all night on the plant with the middle vein severed by a cut. The starch with which the leaf was filled on the previous evening has been removed during the night from all parts of the leaf save those served by the severed vein. (*b*, after Palladin; *c*, after Blue and Stevens.)

chloroplast. This immediate starch formation is utilised in many simple experiments designed to shew the major factors affecting photosynthesis.

(1) *The necessity for light.* A large leaf, still attached to the plant, is kept in the dark until free of starch and is then exposed to light only through the letter spaces of a

stencil which is cut in black paper and fastened over it. After a prolonged exposure, the leaf is detached, decolourised and tested with iodine. The starch formed abundantly below the stencil letters will shew black on a white background the exact shapes cut in the stencil (Fig. 10).

(2) *The necessity for chlorophyll.* A variegated leaf such as *Acer negundo* is allowed to photosynthesise and a sketch is made of the leaf to shew the green and colourless areas. It is then stained with iodine as before, and it will be found that only in the green areas has starch been formed.

(3) *The necessity for carbon dioxide.* A leaf still on a growing plant is enclosed in a glass jar, which is carefully sealed round the leaf stalk by a split cork and wax, or similar device. The jar contains a small tube of caustic soda which robs the enclosed air of all carbon dioxide. After lengthy exposure to the light this leaf and a control from another part of the plant are gathered and tested for starch. Only the latter will be found to contain any, especially if the plant had been somewhat destarched by darkening before beginning the experiment (see also Fig. 10, *b*).

Starch formation. The formation of starch in the chloroplasts is a process which must be most carefully separated from photosynthesis in any theoretical consideration of the processes. The former is a condensation from sugar which requires very little energy uptake and which can go on readily in the dark, whilst the latter is a photochemical reaction involving very large energy absorption. The starch-forming capacity of the chloroplasts (Fig. 11) is shared by the colourless leucoplasts of storage organs such as potato tubers. The ready starch formation which takes place in leaves which are floated on sugar solutions in the dark, indicates that the chloroplasts do not need to make their own sugars as a preliminary to starch manufacture. This experiment is of further interest as shewing that many sugars other

than glucose can be utilised in this way by the chloroplasts.
In the hydrolysis-condensation reaction

$$n\mathrm{C_6H_{12}O_6} \underset{\text{hydrolysis}}{\overset{\text{condensation}}{\rightleftarrows}} (\mathrm{C_6H_{10}O_5})_n + n\mathrm{H_2O}$$
sugar   starch

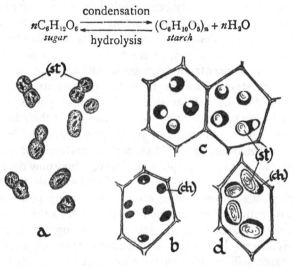

Fig. 11. Starch formation in the chloroplasts. *a, Funaria*, chloroplasts from
the leaf, some of them in process of division. They have been placed in
Schimper's solution (Iodine in chloral hydrate) which has made them transparent, and has stained black the small starch grains (*st*) inside them. *b, c, d*,
cells from the stem of *Pellionia*, shewing stages in the formation of starch. In
the later stages the large starch grain (*st*) has burst out of the chloroplast (*ch*).
(*a*, after Strasburger.)

between sugar and starch, various equilibrium positions seem
to be possible, according to the conditions. Thus at a given
time, in the leaf of a given plant, there seems to be a definite
"critical concentration" of sugar at which equilibrium is possible. If the sugar concentration should rise above this value
(as by rapid photosynthesis) sugar will be condensed to starch
until the critical concentration is reached, and conversely
if the sugar concentration should fall (as by respiration in
the absence of photosynthesis) starch would be hydrolysed
until the sugar concentration again reached the critical value.

Such a phenomenon as this has a very large and interesting number of results in the physiology of the plant. Some mention of it has already been made in the starch-forming and vacuolation regions behind the stem and root apex. The starch-forming tendency is due to a low critical concentration of sugar in the cells and, as the cell ages, the change to a high critical concentration brings about starch hydrolysis and the production of the high sugar concentrations responsible for the cell extension. Different plants differ greatly in the typical critical concentrations of their leaves. Thus the onion and the plants with parallel-veined leaves in general (monocotyledons) have high critical concentrations and only form starch exceptionally, though they may develop very high concentrations of sugar. On the whole the net-veined plants (dicotyledons) are starch-forming plants and have low critical concentrations. Some, such as the tobacco plant, make starch from the first traces of sugar formed in photosynthesis. In a typical starch-forming plant, during the day, starch will be formed by photosynthesis much more rapidly than it can be removed by the combined action of respiration, and the constant diffusion along the veins to other parts of the plant where lower sugar concentrations prevail. (These low concentrations may be due either to direct sugar consumption or to the maintenance of a low critical concentration.) The sugar concentration in the leaf will rise to the critical value and starch will be deposited, so that the granules of it inside the chloroplasts will increase in size. At night, in the absence of photosynthesis, the sugar concentration will be lowered by respiration and translocation out of the leaf, and as it falls to the critical concentration, starch will be hydrolysed until no more remains. Day by day this cycle will be repeated. Thus it follows that if three leaves are cut from such a plant growing in natural summer conditions, one at dawn, one at midday and one at sunset, these leaves will shew vastly different starch content when tested with iodine. The first will contain little or no starch, the second a good deal and the third will be black with it.

The reality of the removal of carbohydrate from the leaf by translocation at night can be easily illustrated by detaching one leaf from a plant at sunset and leaving it all night in a damp vessel beside the parent plant. A starch test of this leaf will shew it to be almost as full of starch in the morning as when it was gathered, but leaves left on the plant will be found to have become empty of starch. By cutting across one or two of the main veins of a leaf at sunset it may similarly be shewn that these are the paths of the sugar translocation, for next morning the areas served by these veins will be found to be still full of starch whilst the rest of the leaf has been quite depleted of it (Fig. 10, c).

The sugar formed in photosynthesis thus has a number of possible fates. It may be directly used in respiration or built into a temporary starch reserve in the leaf, but failing use in the leaf it is moved away to storage organs where it forms carbohydrate reserves of some kind, such as starch in the potato and inulin in the artichoke. It may similarly form carbohydrate reserves in seeds, as starch in maize or wheat, or as reserve cellulose in the date. It may move to a growing region of the plant and there be condensed into cellulose cell-wall, it may form starch and be again hydrolysed, playing its part in cell extension, or it may be built up into the protoplasm of the newly formed cells. The oxygen produced in photosynthesis has a less complex fate, for as it accumulates in the photosynthetic cells, it diffuses from them and goes out of solution at the wet surface of the cells into the air spaces of the leaf. From these, it diffuses through the pores (stomata) in the epidermal layers of the leaf, into the outer air. So long as photosynthesis goes on, the oxygen concentration in the green cells will be kept high and oxygen

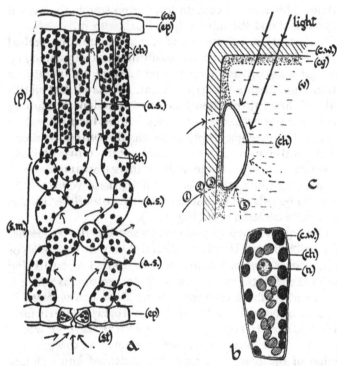

Fig. 12. The passage of carbon dioxide through the leaf to the chloroplasts.
a, small part of a transverse section of a leaf shewing the palisade (*p*) and spongy
mesophyll (*s.m.*) both made up of cells containing chloroplasts (*ch*), and containing air-spaces (*a.s.*). The epidermis (*ep*) on each surface is covered by cuticle
(*cu*), save where the stomatal openings occur (*st*). The arrows shew the paths
of diffusion of carbon dioxide. *b*, single palisade cell; (*c.w.*), cellulose cell-wall;
(*n*), nucleus. *c*, diagrammatic section through one chloroplast lying on the wall
of a palisade cell; (*cy*), cytoplasm; (*v*), vacuole. The numbered arrows shew
the diffusion of carbon dioxide to the lighted chloroplast. Note that part of the
path of diffusion is gas diffusion through the air-space (1); at (2) there is solution
of carbon dioxide in the water in the wet cell-walls; at (3) there is diffusion
through a watery medium of vacuole, cell-wall or cytoplasm. (For leaf-structure
see also Figs. 63 and 65.)

will diffuse to the outer air in which it is present only to the extent of 21 per cent.

Magnitude of the process. Since the photosynthetic process is of such fundamental importance both to the plant and to the rest of the organic world it is of interest to know the magnitude of this process, the means by which the rate of it can be measured, and the factors by which it is controlled. To consider the latter first, we may divide the controlling factors into the internal and external ones. The internal factors include the leaf area, the amount and distribution of chlorophyll, the abundance and shape of the stomata and all such factors as may affect the path of the carbon dioxide or of light to the chloroplast surface from outside the leaf (Fig. 12). Further consideration of these internal factors will be given in Chapter XIV; at this stage we may note that the external factors of light, carbon dioxide concentration and temperature may all play a very important controlling part in photosynthesis. Thus it is clear that in the total absence of carbon dioxide or light, or at low temperature, no reduction of carbon dioxide will occur, and if one factor of these alone is very small it will almost entirely control the rate of the process, however favourable the other factors may be. Thus in a very shady room, photosynthesis will remain very slow however high the temperature and however high the amounts of carbon dioxide in the air. Conversely, if the carbon dioxide is present in very small amounts, increased light will not have much effect on photosynthesis. This has some practical bearing, for if we wish to put up the rate of photosynthesis of a crop of plants we must determine and alter the factor which is chiefly controlling the rate of the process. In nature it turns out that plants are fairly closely adjusted to their environment. On very dull days, the poor light intensity controls the rate of the process, but at most times the small concentrations of carbon dioxide in the air control it and so only by altering this factor could the rate of photosynthesis of a whole crop be increased. This amounts to "manuring" a crop with carbon dioxide, and such a process has actually been used in Germany. The gases from blast furnaces, after purification, are led through pipes and released at soil level. The increased carbon dioxide content of the air round the crops definitely does induce heavier crop yields, although only at a rather heavy cost.

The actual maximal rate of photosynthesis which has been recorded under natural conditions is the following:

Sunflower (*Helianthus annuus*)

28° C.   Sachs 1884   16·5 mg. of carbon dioxide per square decimetre of leaf area per hour.

Later work has served only to confirm Sachs' figures.

By increasing the carbon dioxide concentration to 6·3 per cent. the artichoke (*Helianthus tuberosus*) has been made to assimilate at 30°C. 58 mg. of carbon dioxide per square decimetre per hour, and by increasing both light and carbon dioxide a value of 80 mg. has been obtained for sunflower leaves at 25° C. Plants naturally vary very greatly in the amount of photosynthesis they maintain under the same conditions. In this way plants habitually found in shady places are more efficient photosynthetically in low light intensities than plants which grow exposed to full sunlight. At the maximal rates of photosynthesis recorded for plants growing in the open air, each leaf must, during the hour's photosynthesis, remove the carbon dioxide from a volume of air equal to a column of air the width of the leaf and no less than ten feet high. The photosynthesis of a crop of artichokes has been estimated to produce every year no less than 5½ tons of starch per acre, corresponding with the photo-reduction of 2½ tons of carbon. Any means of putting up these figures is likely to be of the greatest practical

utility to the world at large, either as augmenting the supply
of food or fuel or possibly of oxygen.

The gas exchanges involved in photosynthesis are the
opposite of those in respiration, and by diminishing the

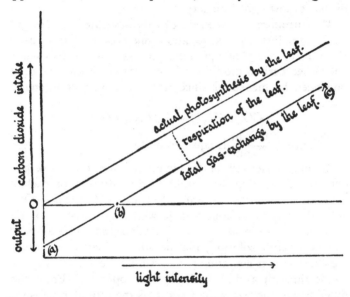

Fig. 13. Gas exchanges of a leaf in different light intensities. (*a*), complete
darkness—no photosynthesis, carbon dioxide output represents full amount of
respiration; (*b*), critical low light intensity—no carbon dioxide given out or taken
in; (*c*), higher light intensities—total carbon dioxide taken in by the leaf
(apparent photosynthesis) equals real photosynthetic intake minus the (much
smaller) output in respiration.

light intensity we can so reduce photosynthesis that a
neutral point is reached at which photosynthesis and
respiration are equal and there is neither intake nor output
of oxygen or carbon dioxide. By diminishing the light still
more, an intake of oxygen and an output of carbon dioxide
is obtained amounting to the full respiration value in com-

plete darkness, but by increasing the light an increasing uptake of carbon dioxide and output of oxygen is apparent. This can readily be seen in Fig. 13, in which intake of carbon dioxide is plotted against light intensity under conditions of low light intensity.

Measurements of the rate of Photosynthesis. The gas exchanges just referred to, afford one means of measuring the rate of the photosynthetic process. Of the two gases concerned carbon dioxide is the easier to measure, and only one need be measured since it is clear from the general equation

$$6CO_2 + 6H_2O \rightarrow C_6H_{12}O_6 + 6O_2$$

that the ratio $\dfrac{CO_2 \text{ in}}{O_2 \text{ out}}$ is equal to unity.

A simple method is to obtain a current from a gas cylinder containing the desired concentration of carbon dioxide, and split the current into two equal streams. One is led directly to an apparatus in which the carbon dioxide is absorbed and measured (e.g. potash or soda-lime tubes or standard baryta solution) and the other passes first through a glass chamber containing the assimilating leaf, and afterwards through a similar absorption apparatus. From the difference in the two sets of readings the rate of photosynthesis can be calculated, and the leaf in the assimilation-chamber can be kept at a constant temperature by immersing the whole chamber in a water-bath, while by using an artificial source, the light-intensity can be varied at will. The oxygen content of the gas streams might also be measured by suitable gas analysis apparatus.

The disadvantage of this experimental method is that it usually involves the use of a cut leaf, and that it cannot be used to measure photosynthesis under natural conditions. In these respects the modified method of Sachs is better. Many large leaves are chosen and, by a rubber stamp, areas of similar size are marked out on either side of the midrib in each one. These areas are chosen to be as much alike as possible. Then, at zero time, the areas are all cut out carefully from one half of each leaf, they are quickly dried at 100° C. and are weighed. The remaining half-leaves are left upon the plant under the given photosynthetic conditions and at the end of the experimental period the areas from these also are excised, dried and weighed. The difference in the weights of the two sets of leaf-areas is reckoned as starch produced by photosynthesis, and may be stated as such, as carbon dioxide or as other convenient units. The translocation which has been going on all through the experiment can be roughly allowed for. Of course it is strictly true that respiration is going on all the time alongside the photosynthesis, so that the results give the "apparent" and not the "real" values for that process. Nevertheless not only does the photosynthesis far exceed the respiration in magnitude, but it is the "apparent" photosynthesis which represents the total gain to the plant and this is the value which, having regard to crop-yield etc., it is most often desirable to measure.

Protein synthesis. The view now generally accepted is that the leaves are the chief centres of protein formation, although other parts of the plant are capable of carrying out the same process at a much slower rate. If nitrate solutions are absorbed by cut leaves it can be shewn that in the light the nitrates swiftly disappear and corresponding extra amounts of oxygen given out by the leaves suggest that they have been reduced to nitrites. The nitrites so formed have been supposed to react at once with the formaldehyde present as a result of photosynthesis in the leaf, to give rise by hypothetical intermediate compounds, to the amino-acids from which the proteins are built up. The actual details of the process are as yet unknown, but it seems clear that we are here concerned with another process which is partly at least photosynthetic. Some of the amino-acids produced are directly built into proteins in the leaf but the majority of them are translocated to meristems and storage organs before such synthesis takes place.

This protein synthesis in the plant is overshadowed by the far greater carbohydrate synthesis, a fact reflected in the utilisation of carbohydrates in all plant activities and plant structures. These seem to be the really abundant substances: the cell-walls of cellulose and lignin are mainly built of them and the respiration is mainly a carbohydrate respiration. The substances which above all others are difficult to obtain by the plant are those involved in the protein supply, and the dilution of nitrates in the soil may lie behind this. Most plants react swiftly to the application of nitrogenous manures and the use of protein in the plant seems most economical. In the animal and in the fungi, both of which live on organic material, this protein shortage is not apparent, large amounts of nitrogenous matter are constantly excreted by the animal, and it has protective and supporting tissue of protein, such as horn, hide, hair and connective tissue, whilst the cell-wall in many of the fungi is a modified form of polysaccharide containing nitrogen in an amino-group on the carbon-atom next to the chain linkage: it is called "fungus cellulose" or "chitin," although apparently not identical with the chitin found in the animal kingdom. The monopoly of effective carbohydrate synthesis on a large scale by the plant world and our complete depend-ance on this process make further investigation of the photosynthetic process a matter of the greatest interest and widest value to the whole of humanity.

CHAPTER VII

YEASTS AND THE BACTERIA

YEASTS

We have already distinguished two types of nutrition, the *holozoic* and the *holophytic*: the former consisting of the ingestion of solid organic food is typical of animals, and the latter consisting of the intake solely of dissolved inorganic material is typical of the green plant. To these must be added the *saprophytic* and *parasitic* types of nutrition. The saprophytic organism absorbs organic food material from the dead body, or the decomposition products of the dead body, of some other organism, either plant or animal, whilst the parasite derives organic food material from the living body of some organism referred to as the "host." The two cases are not altogether separable, for the organic substances, absorbed in liquid form, are usually of much the same kind whether they come from a living or dead body, and it very often happens that a fungus having lived parasitically for a time upon a plant and killed it, may continue to live saprophytically upon its dead tissues. This is the case with the big bracket fungus (*Polyporus squamosus*) which is very frequently to be seen growing on elm trees in this country. Both plants and animals may behave as parasites and each may live upon the tissues of a member of the other type of organism. Thus some fungi, such as ringworm, parasitise animals, and others, like penicillium, parasitise plants. Nematode worms belong to a class of animals parasitic both on plants and on animals. Of the organisms which feed on dead organic matter the plants are said to be saprophytic, and the animals saprozoic.

Both saprophytes and parasites resemble animals rather than plants in that they consume organic material and are generally devoid of chlorophyll and lack photosynthetic

powers. In the plant kingdom almost all their representatives are to be found in the class of the Fungi of which a
very simple representative is the yeast plant, *Saccharomyces*.
The bacteria, which also include parasites and saprophytes,
are best considered as neither plants nor animals (see p. 122).

Structure and occurrence. The yeast plant is unicellular,
and the single cells are about 10μ in diameter, which is
rather smaller than the cells of *Chlamydomonas* or *Euglena*.
The figures (14, *a–f*) illustrate the cell-structure, which can
be made out only with difficulty even under the highest
possible magnifications. The chitin cell-wall encloses a
granular cytoplasm in which a vacuole is fairly conspicuous.
To one side of this vacuole there is a granular dark-staining
body, the nucleus, and strands of chromatin also extend
round the vacuole. The relation of nucleus, vacuole and
cytoplasm is no different from that of other plants, despite
contrary opinions expressed formerly. The extremely small
size of the nucleus makes it very difficult to make out the
details of cell division, but sometimes two chromosomes
are visible, sometimes four. Recent breeding experiments
with yeasts have moreover shewn a genetic behaviour
so like that of higher organisms that a chromosomal
mechanism can be presumed. Within the cytoplasm,
granules or droplets of various substances can be seen varying
in amount according to the metabolic activity of the cell.
Prominent among these are fats and glycogen (a polysaccharide of the starch type), which occur as reserves, the
glycogen at times amounting to as much as 30 per cent. of
the weight of the cell.

Yeasts are to be found growing "wild" in all kinds of
situations in which liquid organic matter is present. Thus
many species of yeast live in the soil upon the broken
surfaces of over-ripe fruit, and in the "slime-fluxes" formed
where branches have been cut from trees. Very few of the
yeasts are parasites, but it is interesting to recall that
one species of yeast lives in the intestine of a small

crustacean, *Daphnia*, and it was observation of the destruction of these yeast cells by the white corpuscles of the
Daphnia which led to the first recognition of the widespread
process of phagocytosis[1] in animals. The yeasts are, however, more familiar as saprophytes, especially in the fermentations of sugary liquids, involved in the manufacture
of alcoholic drinks. In the case of wine and cider-making,
the wild yeasts present on the surface of the ripe fruit may
be used to promote fermentation, but the brewer's yeast
used in making beer is a domesticated yeast not known in
a wild state. In these cases, the use of the yeast depends
upon its capacity for producing alcohol from sugar, and in
breadmaking yeast is utilised because it produces carbon
dioxide which makes bubbles in the dough and causes it to
rise. The significance of these processes to the yeast plant
we shall discuss later.

As is shewn by the variety of different circumstances in
which they will live, yeasts can exist upon a variety of
organic substances, and they will grow satisfactorily upon a
solution of potassium hydrogen phosphate [K_2HPO_4], magnesium sulphate [$MgSO_4$], calcium phosphate [$Ca(PO_4)_2$],
together with ammonium tartrate [$(NH_4)_2C_4H_4O_6$] as the
only organic substance present. The oxidation of the organic salt must yield all the energy and all the nitrogenous
substances necessary for the yeast metabolism, the other
requisites for building the body being available in the inorganic salts. If, to such a solution as this quoted, sugar
(glucose) is added, growth takes place with greater rapidity,
owing to the far greater energy made available to the yeast
by the oxidation of the glucose. By placing a single cell of
yeast into a sterile vessel of such a solution as this, it is
possible to raise an absolutely pure culture of yeast, all

[1] Phagocytosis is the ingestion and breakdown of bacteria by the white blood
corpuscles and other cells of the body tissues.

descendants of one cell; such pure cultures are now kept
and used in the commercial industries of beer and cider-
making, in order that the best strains of yeast may be alone
employed in the fermentation processes.

Reproduction. Under conditions favourable to growth,
yeast reproduces itself by rapid budding. A small part
of the cell-wall softens, and by the turgor of the cell a
bulge is made, which gradually swells until it equals the
size of the parent cell. At the same time cytoplasm passes
into the bud, and the nucleus having divided into two, one
daughter nucleus enters the new cell and one stays in the
old cell. The bud may then be cut off from the parent cell
and commence independent existence. If growth is going
on very rapidly and if the strain of yeast tends to shew
this peculiarity, the first-formed cell may begin itself to bud,
before full sized, and final separation from the parent cell
may be delayed until a whole chain of small yeast cells has
been formed. If the cells bud in more than one place at a
time branched chains of cells will be formed (Fig. 14, a).
Such aggregation is much more usually found in brewer's
yeast than in baker's yeast, probably because the growth
conditions are usually so much more favourable in situations
where the first type grows.

If any yeast is placed under extremely unfavourable conditions, if for example it is starved of organic food by being
spread on a damp block of plaster of Paris, or on the cut
surface of a potato, another method of reproduction will
occur. Under these circumstances the nucleus undergoes
two divisions and the four resultant nuclei become each
enclosed in a round mass of cytoplasm, so that four small
spherical "spores," each with a thick wall, are formed inside
the wall of the parent cell (Fig. 14, f). When this dies and
collapses, the spores are set free and they are so small and
light that they are readily carried away by air currents. The

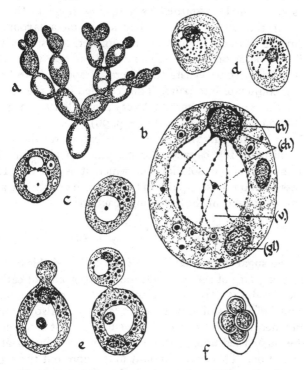

Fig. 14. The yeast plant (*Saccharomyces*). *a*, colony of living cells of brewer's yeast undergoing active budding and forming chains—unstained preparation; *b*, a large scale diagram of a single yeast cell constructed by putting together the information obtained by wide experiments in testing and staining yeast cells with various reagents; (*n*), nucleus; (*v*), vacuole; (*ch*), chromatin; (*gl*), glycogen; *c*, living unstained yeast cells shewing the vacuole and granules of volutin (a reserve substance shewn as black dots), but no nucleus or chromatin; *d*, cells killed and stained to shew the nucleus and network of chromatin round the vacuole; *e*, cells budding; *f*, a cell the contents of which have formed four thick-walled spores. (*a* and *f* after Wettstein, *b*—*e*, after Wager and Penistone.)

contents remain alive for a long time and withstand considerable extremes of dryness and heat and cold, so that the spores can be regarded as a resting stage in the life cycle of the yeast. Should they alight on a favourable medium, containing liquid organic matter, the spores will swell and burst open the wall and bud off new vegetative yeast cells, in some yeasts the liberated spores have been seen to conjugate into pairs. This, coupled with the results of genetic experiment, makes it very probable that in spore production meiois takes place (see p. 96).

Yeast metabolism. The central feature of the yeast metabolism is the capacity for carrying on alcoholic fermentation in sugar solutions in which it is placed. This process goes on most rapidly in the absence of oxygen, and can be expressed in the following reaction:

$$C_6H_{12}O_6 \rightarrow 2CO_2 + 2C_2H_5OH.$$
$$\text{glucose} \qquad\qquad \text{ethyl alcohol}$$

If all the sugar fermented by the yeast followed this course every 100 grams of sugar should yield 48·9 grams of carbon dioxide and 51·1 grams of alcohol, but in practice small amounts of glycerol and succinic acid are also formed. Even when these have been allowed for, about 2 per cent. of the sugar used up cannot be traced in the solution after fermentation. This small amount must represent the sugar used up by the yeast plant itself. This seems a very small amount and it is natural to ask what benefit (if any) the yeast can derive from the process. It is possible that since the yeast can live in alcohol concentrations up to 18 per cent., the alcohol may exclude from the solution all competing organisms to which alcohol is toxic. However, a more significant factor is that the reaction given above is an exothermic one, the fermentation of each gram of glucose setting free energy equivalent to 50 calories. This energy is available, in part at least, to the yeast plant, and though less in quantity than that produced by the complete oxidation of the sugar to carbon dioxide and water, it may be regarded as the key to the meaning of the process. Thus the fermentation is most probably an "anaerobic respiration," that is, a process by which chemical energy becomes available to the yeast metabolism when the yeast is growing in anaerobic conditions, i.e. growing in the absence of oxygen. Similar anaerobic processes are found in other organisms, notably the bacteria, and may even occur in the cells of the higher plants if these are kept devoid of oxygen.

Although capable of a maximal rate of fermentation when kept without oxygen, yet when so treated the activity of the cells diminishes, and they cease to bud and grow. On the other hand, when allowed oxygen, although the production of alcohol falls off considerably, the cells increase in size and multiply very rapidly. Thus, in brewing technique, conditions have to be adjusted so that aeration is given adequate to permit growth of the yeast cells, but not so great as to diminish alcohol production unduly.

Under *aerobic* conditions, it seems likely that three major reactions are taking place:

(1) alcoholic fermentation

$$C_6H_{12}O_6 \rightarrow 2CO_2 + 2C_2H_5OH.$$
$$\text{glucose} \qquad\qquad \text{alcohol}$$

(2) aerobic respiration

$$C_6H_{12}O_6 + 6O_2 \rightarrow 6CO_2 + 6H_2O.$$

(3) glycogen formation

$$n(3C_2H_5OH) + n(3O_2) \rightarrow (C_6H_{10}O_5)_n + n(4H_2O).$$
$$\text{alcohol} \qquad\qquad\qquad \text{glycogen}$$

In the past, it used to be thought that the alcoholic fermentation process was merely the initial stage of the ordinary process of aerobic respiration—a process completed, in the presence of oxygen, by a second phase of oxidation of alcohol to carbon dioxide and water

$$[C_2H_5OH + 3O_2 \rightarrow 2CO_2 + 3H_2O];$$

but arrested in the absence of oxygen at the stage of alcohol formation and accumulation. The modern work which has revealed the existence of rapid "building back" of alcohol to glycogen in the presence of oxygen has so complicated our views as to the nature of the yeast respiration mechanism, that it can no longer be expected to yield such a simple key to the respiration of the higher plants.

Brewing. The processes involved in brewing are of considerable interest, and scientific investigation of them has been very extensive. They begin with malting; fresh barley grains are first soaked, and then kept moist and warm on the floor of a malting house; they are raked over from time to time and soon they begin to germinate. Very extensive changes go on in the grain; abundant supplies of enzymes are produced, and by these the starch reserves are largely converted into sugar (by amylase), and the protein reserves form large amounts of soluble nitrogenous subtances (by proteases). These are so produced in all germinating seeds, the reserves thus becoming available for the respiratory and synthetic processes of the growing seedling. The malt (the germinating grain) is then killed by raising the temperature to about 70° C. and it is dried in large malt kilns. If stout or porter is to be made the malt is charred. The killed malt is ground up and mashed in water at temperatures between 60° and 70° C. This serves to dissolve out soluble matter and to make still more matter soluble. In particular, the hydrolysis of starch is continued and dextrin, maltose, and glucose are produced in relative amounts varying with the temperature. The kind of beer thus depends not only on the materials but on their treatment, and kiln and mash-ing temperatures have to be most carefully controlled. The malt is removed from the mash and the liquid (wort) is boiled with hops for some time. This extracts the bitter aromatic substances from the hops, and sterilises and concentrates the wort and precipitates nitrogenous matter. The liquid is allowed to cool and settle in a shallow vessel of great area, which is usually refrigerated, and then yeast cultures are added and it is allowed to ferment for some days. The yeast cells ferment the sugars in solution, and conditions are so adjusted that they multiply rapidly and yet produce alcohol at a rapid rate. The carbon dioxide produced at the same time causes a great deal of frothing and it may accumulate in dangerous amounts near the top of the vat. The yeast is cleared from the solution by various methods and the beer is run into special storage vessels, after which suspended matter in it is precipitated as far as possible and it is bottled. In order to make the beer bubble attractively when poured out, carbon dioxide under pressure may be forced into it during bottling, a process called carbonisation.

Wine making is similar in principle to brewing but it is less scientifically managed. The "must" from the wine press is the liquid which corresponds to the wort which is fermented, and as before stated the yeasts responsible are the wild ones present on the fruit.

Enzyme content. One of the most interesting features about a study of the yeast plant is that the enzymes which carry out its various activities have been very intensively studied. The first known enzyme was the "zymase" prepared by Buchner in 1897. He ground up the living yeast cells with sand and rotten stone and then pressed out the mass under 400 atmospheres pressure. The yeast juice so obtained was found to be able to carry out the alcoholic fermentation of sugar, though the property was destroyed by exposure to temperatures of 40° to 50° C. From this juice the single enzyme zymase was prepared in a relatively pure state and it appeared to be a white powder which would last indefinitely if kept dry. Buchner's preparation was much less active than the modern preparations of "zymin." This is merely a mass of dead yeast cells, which are first rubbed into acetone through a fine sieve, and are afterwards washed and dried. Such treatment kills the cells, but does not destroy the enzymes, and it permits the enzymes to diffuse through the cell walls as they cannot do in the living cell. Since the yeast cell contains a full outfit of the protoplasmic enzymes, zymin is a ready source of all kinds of enzymes. Besides zymase it contains lipase, proteases, the carbohydrate enzymes and oxidative and reducing enzymes. In the living cell, these, with the exception of one, are all bound to the protoplasm so that they carry on reactions only inside the cell, but the exception, invertase, diffuses out of the living yeast cell and will split up cane-sugar into dextrose and laevulose in the absence of the yeast cell. This is a subsidiary reaction often employed in brewing practice; cane-sugar which has been acted on by yeast is added to the mash and the yeast reaction thus increases the amount of fermentable sugar in the wort. Zymin has been used commercially but not very satisfactorily since it loses its activity after two or three days, is of course very sensitive to high temperatures, and is non-regenerative.

Progressive biochemical work on the yeast enzyme system has greatly increased our knowledge of enzyme behaviour in general, and seems at present our most favourable approach to an interpretation of the fundamental process of plant respiration in terms of enzyme activity.

BACTERIA

The bacteria constitute a small group of organisms remarkable for their small size and for the variety and importance of their activities in all parts of the organic world. The average-sized bacterial cell is only about $1\,\mu$ in diameter, and organisms are known to exist which are small enough to be invisible below the microscope and to pass through porcelain filters. Such organisms are responsible for diseases such as swine fever and foot and mouth disease, and some idea of their size can be gauged by determining whether they will or will not pass through different members of a series of collodion filters prepared so as to have pores of different sizes. By these means these "viruses" are found to consist of units of about $0.1\,\mu$ in diameter. Although a class of bacteria exists which includes fairly large filamentous forms, the majority are simple unicellular units, and it is with these that we shall be especially concerned. In the early days of bacteriology, the bacteria were classified into various "form-types" as below. (1) The *bacillus*, a straight rod-like cell, cylindrical and much longer than wide; (2) the *coccus*, completely spherical; (3) the *spirillum*, a curved rod. By prefixes the manner of aggregation of the cells could also be indicated—thus *Streptococcus* meant cocci arranged in chains, and *Staphylococcus* meant cocci arranged in close bunches (Fig. 15). The numbers of such names grew extremely large, but even so the system proved inadequate as a means of distinguishing one bacterium from another, and bacteria absolutely alike in form were found to be perfectly distinct organisms. Thus *Bacillus typhosus*, which produces typhoid fever, very closely resembles the harmless *B. coli communis* which is always present in the human intestine though they are readily separable by the chemical changes they produce in the various media upon which they may be grown.

It is natural that in so small a body as the bacterial cell

little differentiation should be observable. There is a cell-wall or cell-surface which is very thin and rather like that of an animal cell. In addition to this there may be present a much thicker wall, the capsule, which is composed in part at least of carbohydrates and which is often so mucilaginous that it causes the bacteria to cling together in so-called "zoogloea masses." There is a central structure within the cells of some bacteria which takes up nuclear stains, and this body divides just before cell-division, as a nucleus would do. In some species of bacteria small protoplasmic threads protrude from the surface of the cell and by the movement of these cilia or flagella the bacterial cell swims actively in the liquid medium. Other motile bacteria shew the property of response to stimuli, in that they swim in a definite manner to or away from the source of some stimulus. Thus *Bacterium termo* has already been mentioned as sensitive to free oxygen, and other bacteria are sensitive to the presence of other chemical substances in solution. As the bacteria respond by movement of the whole organism the response is said to be a chemo*taxis* (see p. 57).

Reproduction. *Fission.* When placed in suitable conditions the bacterial cell multiplies, usually by the process of simple fission. The cell becomes constricted across the middle and the protoplasm divides into two portions, each of which forms a new cell. If the cell membranes tend not to cohere together, the cells separate, but in some kinds of bacteria the membranes do not readily separate, so that the dividing cells become aggregated in the mucilaginous *zoogloea* masses referred to above. In favourable situations a bacterium cell may become mature and able itself to divide within twenty minutes of its origin by fission from a parent cell. Such rapid multiplication would give in twenty-four hours a progeny of several millions from each bacterial cell, but this rate of reproduction would only be possible where the nutrient medium was adequate for all the cells produced, and in practice this limit is soon reached, so that we may imagine that everywhere the bacterial population is closely limited by the food supply available. Temperature also naturally has a very great effect upon the rate of bacterial reproduction, and different bacteria have different optimal temperatures for their growth. Thus the bacteria which inhabit the human body flourish best at 37° C., there are soil bacteria with optima from 20° to 24° C., and bacteria of dungheaps with optima about 70° C.

Other factors which affect bacterial growth could be grouped under the heading of food supply. Thus the precise type of organic and inorganic substances available is important, and the supply of gaseous oxygen and of water all are controlling factors of the utmost importance. As the optimal amounts of each differ from one species to another, they afford a means of differentiating the various bacteria from each other. In this way the bacteria may be divided into obligatory aerobes—those which will only live in the presence of free oxygen, obligatory anaerobes—those which will only live in the complete absence of gaseous oxygen, and facultative anaerobes—those to which the presence or absence of free oxygen seems not to matter. *Clostridium tetani*, which is responsible for tetanus, is an obligate anaerobe, and when grown in cultures such a bacterium will develop only in those parts of the medium furthest away from the free air surface. On the other hand, the hay-bacillus (*B. subtilis*) will only grow at the surface since it is an obligate aerobe. This relationship to the oxygen supply has, as we shall see later, a direct connection with the chemical life-processes of the bacteria.

Spore formation. As in the yeast plant, bacteria, especially those of the bacillus form, may produce *spores*, on the incidence of certain conditions, usually unfavourable ones. The protoplasm inside the bacterial cell produces a single, very thick-walled protoplasmic cell, usually round or oval, and quite non-motile. The spore, lying within the old parent cell, gives a strikingly characteristic appearance to many different kinds of bacteria, e.g. *B. anthracis* (see Fig. 15, *f*).

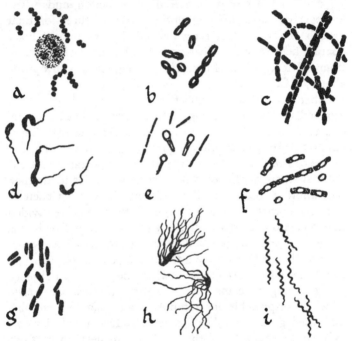

Fig. 15. Bacteria. *a*, Streptococci from pus (shewing also a blood corpuscle); *b*, *Pneumococcus* (pneumonia); *c*, *Bacillus anthracis* (anthrax) shewing bacilli in chains; *f*, *B. anthracis* shewing the formation of a single spore inside each bacterial cell; *d*, cholera spirillum shewing flagellate cells; *e*, *Clostridium tetani* (tetanus) shewing cells with and without spores; *g*, *h*, *Bacillus typhosus* (typhoid) with and without flagella; *i*, *Spirochaete pallida* (syphilis).

The spores so formed possess, partly no doubt on account of their thick wall, extraordinary powers of resistance. Thus Koch, the great German bacteriologist, found that whilst ordinary vegetative cells of *Bacillus anthracis* were killed in two minutes by 1 per cent. carbolic acid, the spores would resist the same solution for as long as 15 days. Some spores will resist a dry heat of 100° C. for a long time, and may even withstand boiling. Upon reaching a suitable moist medium the spore grows out again into the ordinary vegetative form of bacillus.

Dispersal and Sterilisation. The distribution of bacteria, like that of the yeasts, is largely a matter of wind carriage. They are so minute that ordinary vegetative cells and spores alike remain suspended in the air by the faintest air currents and they are usually present in the air, though in smaller numbers than is usually thought. The rapid commencement of souring of milk and beer, and the rotting of organic bodies may be due to infection from the air but is more often due to more indirect infection.

Preservation from bacterial infection and attack usually involves sterilisation of the material to be preserved, and the necessary killing of all bacterial life in a body can be achieved in various ways. A dry heat of 160° C. for half an hour or 180° C. for ten minutes is effective, but it is more usual to use steam at high pressures, and with this technique ten minutes at 115° C., or five minutes at 120° C., will kill the most resistant spores. If simple boiling is used, the process must be repeated three times on each lot of material in order to destroy the progeny of spores not killed at the first or second attempts. Filtering through unglazed earthenware has been used as a method of freeing drinking water from bacteria, but the method is uncertain for more than a rough precaution. Bacteria are fairly readily killed by a whole range of chemical substances called antiseptics. The best known include iodine, alcohol, mercuric chloride, potassium permanganate, carbolic acid, hydrogen peroxide, formalin, iodoform, chlorine (used as bleaching powder), and sulphur dioxide. In addition to these, the germicidal effect of direct sunlight is now recognised, typhoid bacilli and dry anthrax spores both being killed by $1\frac{1}{2}$ hours sunlight. It has been determined that the ultra-violet rays are the most effective components of the sunlight for such purpose, though the rest are not ineffective. The introduction of heat sterilisation methods by Pasteur and the use of antiseptics by Lord Lister marked tremendous advances in our control of bacterial activities.

Cultivation of bacteria. The most useful technique now known to the bacteriologist is that which enables him to grow each bacterium alone in a test-tube on an artificial medium uncontaminated by other organisms, and under closely controlled conditions. This is known as growth in pure culture, and in order to achieve it various types of difficulty have to be met. In the first place the medium for the culture must be completely sterilised; this is possible by closing the containing vessel (flask or test-tube) with a plug of cotton wool and sterilising by heat in the ways already mentioned. The cotton wool serves as a filter which admits air, but keeps out bacteria when the air cools and the air pressure inside the vessel falls. The requisite food media have to be determined, and those selected which will not be destroyed by the sterilisation process. Finally, the bacteria have to be isolated from the complexity of the situation in which they occur, often many million to the cubic centimetre, and mixed up together one species with another. This last difficulty is resolved by the method of "plating out" introduced by Koch. The bacteria are not allowed free play in a liquid, but are rendered immobile by growth in a gelatinous medium. A portion of the bacterial mass to be investigated is spread over the surface of the medium, discrete colonies of the bacteria develop, and these may then be examined directly or isolated and grown in separate cultures. In making a culture they are transferred to a sterile test-tube of medium by a needle, and in the tube they form colonies often of striking and characteristic shape and colour, varying with the nature of medium and of the bacterium. The growth of a given unknown bacterium on a range of media, such as gelatine, agar, meat-extracts, blood serum, potatoes, lactose, milk and peptone, gives data of shape and size and colour and growth rate of the colony, of reaction to oxygen and of reactions on the medium, all of which assist an accurate "bacterial diagnosis." Other valuable tests in the identification of bacteria depend on the reaction of the bacteria to various stains, and especially to their capacity or incapacity to retain the stains after treatment with alcohol or mineral acids. The staining is carried out on a slide and the bacteria are then examined microscopically. The most accurate means of identification depends on the presence in the bacteria of specific protein bodies known as "antigens." When a culture of bacteria is introduced into an animal body the antigens cause the production there of corresponding protein substances, the antibodies. When serum from the animal is added to the original bacterial culture the antibodies cause precipitation or agglutination reactions visible in the test-tube. The serum will not so react with bacteria other than the original type. By this method it is possible to match an unknown bacterium against a set of known cultures, such, for example, as those of the different forms of typhoid.

Natural occurrence of bacteria. The circumstances in which bacteria can grow include the variety of food material which they can utilise, and so depend clearly upon the reactions of their metabolism. Only when this metabolism has been fully considered shall we be able to see the significance *to the bacteria* of the variety of circumstances in which they grow, but the significance of their activities *to us* may be considered now.

The first significance seems to lie in the connection of bacteria with diseases both of plants and animals. In both cases the bacterial action may arise in various ways. In the human body the pathogenic bacteria produce their ill-effects sometimes by the direct breakdown of the body tissue, and sometimes by the secretion into the blood of toxins which affect physiological activities, and cause tissue destruction at a distance from the bacteria. The biology of such pathogenic forms is a subject of the widest possible interest. In some cases the diseases are "infectious," and the bacteria are transferred directly or indirectly from one person to another. Such are tuberculosis (*Mycobacterium tuberculosis*), syphilis (*Spirochaete pallida*), typhoid (*B. typhosus*), etc. In some cases the pathogenic bacterium may be a normal inhabitant of soil or dung and becomes dangerous on entering a wound in the body. An example of this is *Clostridium tetani*, responsible for tetanus. Alongside the pathogenic forms there exist in the body large numbers of bacteria which are harmless, or even useful to the animal. All these facts have their bearing on the nature of treatment and investigation of bacterial disease.

Not by any means are all bacteria harmful. A very large part is played by bacteria in bringing about the putrefaction of dead organisms either animal or plant in nature. The bacteria exist as saprophytes on such bodies and break down their tissue structure, and reduce their chemical substances to simpler forms from which bacterial protoplasm is made. On the death of the bacteria these substances become readily available in the soil to green plants, and again play a part in the synthesis of complex organic matter suitable for animal food. Some bacteria have the power of "fixing" atmospheric nitrogen and building it up into their protoplasmic proteins. When they die, they enrich the soil greatly in respect of its nitrogen content. Especially is this true of a bacterium (*Bacillus radicicola*)[1] which grows in nodules on the roots of plants of the pea, bean and vetch family. The growth of such plants in poor soils often actually increases the soil fertility, whilst most other plants constantly deplete the soil of nitrogen.

[1] Now more accurately called *Rhizobium leguminosarum.*

Still other bacteria have important industrial uses, and play a part in the curing of tobacco, the "retting" of flax (the rotting of the softer tissues from the useful fibres), the ripening of cheese, and innumerable other processes where we seldom suspect their activities. Putrefactions, the decay of food material, and even the decay of fabrics, are often bacterial in origin. The significance of the *rôle* the bacteria play in the life of the world is extraordinarily wide; nevertheless we shall be in error if we consider their activities only from the human point of view. For the biologist all these activities must also have their interpretation in the terms of the physiology of the bacteria themselves; we must attempt to envisage them "from the point of view of the bacterium," that is, as part of the active bacterial metabolism.

The bacterial metabolism. The nutritional problems with which all plants, and indeed all animals, are faced have a twofold aspect. The material intake of the organism must supply (1) the raw substances out of which new protoplasm and skeleton, that is growing organism itself, can be constructed. Such substances we have shewn, in Chapter IV, to be inorganic for the green plant, and organic for the animal. The material intake must also supply (2) some source of energy to be the motive power of the metabolism of the organism. This source of energy we have seen to be, in the green plant, ultimately the process of photosynthesis, with its intake of carbon dioxide and light energy, although the energy is released and made available by the oxidation of carbohydrates during respiration.

In all organisms such a dual necessity as this must exist although possibly one type of material taken in by the organism may supply both needs. Thus the organic matter digested by the animal serves both as food material by which its body substance is synthesised and as respiratory material liberating energy on oxidation. Similarly, the photosynthetic process in the green plant supplies all the

material for the respiratory processes and also supplies some
of the material needed in the synthesis of the plant body,
though not by any means the whole demand, for a large
intake of mineral salts is also necessary. As may be seen
from the table opposite, there are organisms, such as the iron
and sulphur bacteria, in which the twofold nutritional demand concerns two entirely distinct types of material, that
respired playing no part as the raw material of the body
substance of the organism. The metabolism of such bacteria,
which is of great interest, will be dealt with more fully on
p. 116.

In those bacteria in which the raw material of the body
substance is taken in as inorganic or simple organic substances, a great deal of energy will be required to complete
this ultimate synthesis. Especially large amounts of energy
will have to be utilised in the reduction of carbon dioxide
to carbohydrate, and this will not be achieved by photosynthesis, as in the green plant, but it will come about by
energy derived from the respiratory, oxidative processes.
This synthesis will then be termed *Chemo*-synthesis, as
opposed to photosynthesis, since the energy for it is derived
from an ordinary chemical reaction. Such reduction is by
no means unknown. Fenton has shewn the production
of formaldehyde from the aqueous solution of carbon
dioxide in the presence of metallic magnesium, and the same
reaction can be produced by the silent electric discharge
and by ultra-violet light. The polymerisation of the formaldehyde so formed into sugar has also been carried out
in vitro. There is nothing unusual about the reactions involved in chemosynthesis, except perhaps in the variety of
reactions utilised. It will be clear that the metabolism of a
bacterium will depend as closely on supplies of material
for respiration as upon supplies of material needed for
building up into body substances. Both must be noted in

Organism	A Intake of raw material out of which the body substance is built up.	B Intake of material utilised in process C.	C Reaction by which energy is liberated in the service of metabolism. (Respiration.)
Green plant	Inorganic salts. Water. Carbon dioxide and light (photosynthesis).	Carbon dioxide and light (photosynthesis) forming carbohydrate. Oxygen.	Oxidation $C_6H_{12}O_6 + 6O_2 \rightarrow 6CO_2 + 6H_2O + E.$ glucose
Animal	Organic matter. Proteins, carbohydrates, fats, etc.	Organic matter. Proteins, fats, carbohydrates, etc. Oxygen.	Oxidation of proteins, carbohydrates, fats, etc.
Fungus	do	do	do
Yeast	Simple organic and inorganic salts. Carbohydrates.	Carbohydrates, e.g. glucose. Oxygen.	Anaerobic splitting $C_6H_{12}O_6 \rightarrow 2C_2H_5OH + 2CO_2 + E.$ glucose alcohol Aerobic oxidation $C_6H_{12}O_6 + 6O_2 \rightarrow 6CO_2 + 6H_2O + E.$ Aerobic glycogen formation $n3(C_2H_5OH) + n3(O_2) \rightarrow (C_6H_{10}O_6)_n$ alcohol glycogen $+ n4(H_2O) + E.$
Iron bacteria	Inorganic salts. Water. Carbon dioxide.	Ferrous salts.	Oxidation $Fe'' + O \rightarrow Fe''' + E.$ ferrous salts ferric salts
Sulphur bacteria	do	Sulphuretted hydrogen.	$H_2S + O \rightarrow S + H_2O + E$ sulphuretted sulphur hydrogen $S + O_2 \rightarrow SO_2 + E.$ sulphur dioxide $SO_2 + O \rightarrow SO_3 + E.$ sulphur trioxide

the following examples of different types of bacterial metabolism.

Iron, Sulphur, and Methane Bacteria. A number of bacteria utilise as an essential part of their metabolism, such unlikely material as sulphuretted hydrogen, methane or ferrous iron. For instance the sulphur bacteria of the beggiatoa group live in situations where sulphuretted hydrogen occurs in solution, having been liberated by the decay of organic material containing protein. This substance is taken in by the bacteria and is oxidised by them according to the following equation:

$$2H_2S + O_2 \rightarrow 2H_2O + 2S.$$

The reaction takes place inside the cells and granules of sulphur are deposited within them. If carbon dioxide is also supplied to the bacteria no other organic material is necessary. The energy set free by the sulphuretted hydrogen suffices to bring about reduction of the carbon dioxide and to build up from the simple substances so formed, and from inorganic salts, all the protoplasmic material of the bacterium itself. Thus we see that the general anabolism of the organism is on one line of chemical activity, and the energy supply for this anabolism is derived from an entirely separate line of chemical activity, upon quite different material. Thus a non-carbohydrate respiration takes the place of the ordinary carbohydrate respiration of green plants, and organic material is built not by photosynthesis, but by chemosynthesis.

The respiratory oxidation reaction may go beyond the stage we have indicated; the sulphur deposited in the bacteria may be oxidised to sulphuric acid, especially if neutralising substances such as lime are present. The whole reaction could be written

$$S + 3O + H_2O \rightarrow H_2SO_4 + 141 \text{ cals.}$$

The methane bacteria live in a precisely similar way, but their respiratory oxidation is that of marsh-gas:

$$CH_4 + 2O_2 \rightarrow CO_2 + H_2O + 220 \text{ cals.}$$

In districts where ferruginous rocks are drained by streams, often the water is covered with slimy, bright reddish-brown masses of iron bacteria. These are utilising the oxidation of ferrous to ferric salts as their respiratory process. Other bacteria may even utilise the oxidation of free hydrogen in the same way.

Soil bacteria and the Nitrogen Cycle. The diagram given in Fig. 16 will serve to illustrate the circulation of nitrogen in nature, and the part played in it by bacteria and the green plant.

Dead organisms undergo breakdown by the autolysis produced by their own liberated enzymes, and in this and other ways their complex proteins appear in the simplified form of amino-acids. Such amino-acids are split up, especially by the bacterium *Pseudomonas*, with a production of ammonia which is set free in the soil. This ammonia (combined as carbonates by reaction with carbonic acid in the soil) is acted upon by other bacteria, such as *Nitrosomonas* and *Nitrococcus*, and is oxidised to nitrous acid in the soil (this again forming nitrites at once with bases in the soil). The nitrites undergo further oxidation by the bacterium *Nitrobacter*, and nitrates are produced. These reactions thus form a chain by which the complex animal and plant protein is rendered again available to green plants, for only in the form of nitrates is nitrogen taken in by them in large quantity. Since the green plants synthesise the bulk of the organic protein food of the animal world, the cycle keeps continually in movement, and the importance of the bacteria in it cannot be overlooked. The chain of processes is called "nitrification," and the bacteria are termed "nitrifying bacteria." To the bacteria themselves, the reactions mentioned are all respiratory-oxidation processes, which make possible chemosynthesis and all the anabolic activities of the bacteria. In the table on pp. 120–21 all these reactions are set out, and they will be recognised as all exothermic. Fig. 16 shews that all the nitrifying bacteria utilise small amounts

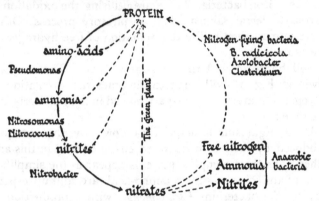

Fig. 16. The nitrogen cycle. A schema to show the parts played by bacteria and green plants in the circulation of nitrogen in nature. The more heavily printed names are those of the chemical substances involved and those in smaller print are the agents bringing about their alteration. The continuous lines indicate reactions in which energy is set free (exothermic) and the broken lines reactions using up energy (endothermic).

of the substances they oxidise, to synthesise their own protoplasmic proteins.

To complete the picture of the nitrogen cycle it is necessary to describe also "denitrification" and "nitrogen fixation." The former process is, as the name implies, the reduction of nitrates to nitrites, ammonia or free nitrogen, and it is a process harmful to the soil fertility, as it lowers the nitrate content. Such a process takes place under anaerobic conditions such as are found in water-logged soils. Under these circumstances anaerobic bacteria develop, and their demand for oxygen is responsible for the reduction of

nitrates and sometimes of sulphates in the soil. In contrast
with the nitrifying bacteria, the reaction upon the nitrogenous compounds of the soil is not significant to the
denitrifying bacteria as a direct oxidation respiration process, for the change is endothermic. The same is true of the

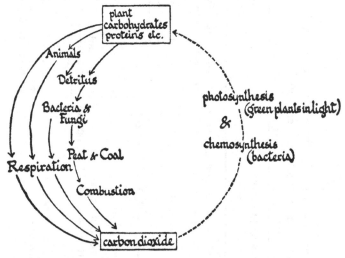

Fig. 17. The carbon cycle. Schema to shew in a general way the parts played
by different types of organisms in the natural circulation of the element carbon.

nitrogen-fixing bacteria. These include several kinds of
bacteria, some aerobic such as *Bacillus radicicola* and *Azoto-
bacter*, and some anaerobic, such as *Clostridium pasteuria-
num*, but in all of them free gaseous nitrogen from the air
is synthesised into bacterial protoplasm in large amounts.
As this process requires energy, its significance to the bacteria
does not lie in the demand of the respiratory system of the
bacterium, but in the demand of the anabolism for nitrogenous material. The energy for the anabolism is produced
by a respiratory oxidation of sugar, such as glucose, derived
from decaying carbohydrate matter, etc. in the soil.

Table to shew the type of reactions carried out by the bacteria involved in the nitrogen cycle, especially in relation to the metabolism of the bacteria themselves.

		EXOTHERMIC PROCESSES Oxidative respiration	ENDOTHERMIC PROCESSES Anabolism
	Sulphur bacteria	Sulphuretted hydrogen or sulphur to sulphuric acid. $S + 3O + H_2O \rightarrow H_2SO_4 + 141$ cals.	$CO_2 \rightarrow$ carbohydrates $\rightarrow$ inorganic salts $\rightarrow$ } bacterial protoplasm
NITRIFYING	*Pseudomonas*	AMINO-ACID TO AMMONIA $NH_2 - CH_2 - C \begin{smallmatrix} O \\ OH \end{smallmatrix} + 3O \rightarrow 2CO_2 + H_2O + NH_3 + 152$ cals. glycine	$CO_2 \rightarrow$ carbohydrates $\rightarrow$ amino-acids $\rightarrow$ proteins $\rightarrow$ inorganic salts $\rightarrow$ } bacterial protoplasm
	Nitrosomonas *Nitrococcus*	AMMONIA TO NITRITES $(NH_4)_2CO_3 + 3O_2 \rightarrow 2HNO_2 + CO_2 + 3H_2O + 148$ cals. ammonium carbonate	$CO_2 \rightarrow$ carbohydrates $\rightarrow$ ammonia $\rightarrow$ proteins $\rightarrow$ inorganic salts $\rightarrow$ } bacterial protoplasm
	Nitrobacter	NITRITES TO NITRATES $KNO_2 + O \rightarrow KNO_3 + 22$ cals. potassium nitrite	$CO_2 \rightarrow$ carbohydrates $\rightarrow$ nitrites $\rightarrow$ proteins $\rightarrow$ inorganic salts $\rightarrow$ } bacterial protoplasm

		OXYGEN FROM NITRATES
DENITRIFYING	Anaerobic splitting of organic material, e.g. cellulose, sugar	nitrates $\left\{\begin{array}{l}\rightarrow \text{nitrites} + O_2 \\ \rightarrow \text{ammonia} + O_2 \\ \rightarrow \text{free nitrogen} + O_2\end{array}\right.$ Oxygen utilised either in building bacterial protoplasm or in oxidation processes.
NITROGEN FIXING — *B. radicicola* *Azotobacter*	Splitting up of organic material, e.g. sugar or cellulose. aerobic $C_6H_{12}O_6 + 6O_2 \rightarrow 6CO_2 + 6H_2O + 677$ cals.	FREE NITROGEN TO $\left\{\begin{array}{l}\text{PROTEINS} \\ \text{carbohydrates} \longrightarrow \\ \text{inorganic salts} \longrightarrow\end{array}\right\}$ bacterial protoplasm
NITROGEN FIXING — *Clostridium pasteurianum*	sugar anaerobic $C_6H_{12}O_6 \rightarrow C_2H_5OH + 2CO_2 + 50$ cals.	

(The reactions forming part of the nitrogen cycle are put in capitals : the sulphur bacteria have been included for comparison ; the whole scheme is intended to illustrate the general trend of the metabolism and is not to be taken as a guide to details ; the equations are given as examples.)

Bacterial breakdown of carbohydrate material plays a large part in the destruction of dead vegetable material in the soil, such as the straw of manure. Such breakdown produces the "humus" of the soil, dark-brown colloid matter which covers the small soil particles, improving the water-holding powers of the soil, its aeration, and general crop-producing capacity. A bacterium of this class is *Spirochaete cytophaga*, which has been recently employed in the production of artificial farmyard manure from straw moistened with ammonium sulphate solution. The ammonium salt supplies the nitrogen necessary for synthesis of the bacterial protoplasm. It is clear that the bacteria play a part in the carbon cycle of nature as they do in the nitrogen cycle, though certainly a less important one. This may be gauged from the diagram of Fig. 17.

It has been recently shewn that several kinds of bacteria have photosynthetic powers: they are anaerobic forms which live just below the surface of organic mud at the bottom of lakes and ponds. Some groups of them are green and others purple. Both contain pigments almost identical with the green chlorophyll pigments of the higher plants, but carotin and xanthophyll are lacking, though sometimes a related pigment, spirilloxanthene, occurs. These bacteria will carry out photoreduction not only of carbon dioxide, but also of nitrate, of many sulphur compounds and of a big variety of organic substances. As they also have powers of chemosynthesis of carbon dioxide through the high energy compounds produced in photosynthesis, they are completely autotrophic organisms. The close resemblance of the pigments to those of the higher plants certainly suggests some evolutionary relationship, and investigation of bacterial photosynthesis may well throw much light on that of the plant kingdom as a whole. Already it suggests that the yellow pigments may not have a primary rôle in the photosynthetic process.

CHAPTER VIII

THE FUNGI

The yeast plant has already been described as a member of the fungi, but although it has the nutrition typical of this group its structure is somewhat simpler. The typical fungus has a body devoid of chlorophyll, and consisting of a colourless branched tube of fungus-cellulose, with or without cross walls. In the protoplasm lining this tube (or hypha) there are numerous nuclei which multiply by division where extension is taking place, at the apices of the hyphae. The system of hyphae which constitutes the body of the fungus is called the "mycelium." In accordance with the nutritional habit of the fungi, which is saprophytic or parasitic, a great part of the mycelium lies in the food material, penetrating it in all directions and absorbing over all its surface organic materials which are already soluble, or which it renders soluble by the secretion of enzymes. Only the reproductive hyphae commonly come above the surface of the nutrient medium. The large surface and extreme permeability to water, which make the fungal hyphae such an effective means of absorption, at the same time make them extremely susceptible to desiccation and it is probably for this reason that fungi flourish best in very moist situations, and may often be killed by dry conditions. Such for instance is the case with the fungus (*Pythium*) which causes the damping-off of seedlings which are over-watered or kept in too moist air (see p. 132).

Mucor. Many organic substances, such as cooked food of various kinds, bread, jam, leather, etc., when left in damp situations speedily become covered with a felt of the mycelium of saprophytic fungi. These are generally white, yellow or blue-green in colour, and are popularly spoken of as "moulds." One of the commonest of them is *Mucor*, which is sometimes called "pin-mould" from the appearance of the reproductive branches.

Mucor grows on a very wide range of organic material; one of the commonest occurrences is on jam. The mycelium on the surface of the jam takes in organic material, chiefly glucose, in solution, and oxygen is taken in from the air. The respiration of the mould seems in most cases to be a simple aerobic oxidation of the glucose, and for this reason the growth is limited to the surface layers of the jam where free oxygen is available. In some species of *Mucor*, however, an anaerobic existence is possible, singularly like that of the yeasts. Not only is the anaerobic respiration the same, involving the formation of alcohol and carbon dioxide from glucose, but the fungus mycelium, when in a liquid medium, often forms small single cells like yeast cells which reproduce by budding. It is the activity of such species of *Mucor* which ferments the deeper layers of jam, causing the appearance of small bubbles of carbon dioxide, and the taste and smell of alcohol. Similar moulds have been employed in a very widespread way in the production of various alcoholic drinks from sugary media.

It is interesting to note that in *Mucor* there is present the same lack of movement, the same branching, and the same indefinite and continuous growth, that we find in the higher plants. As with them the apical region is sensitive to stimuli and will turn and grow towards sources of food material and of oxygen.

Reproduction. (a) *Spore formation.* The most common means of reproduction in *Mucor*, as in the great majority of fungi, is by the production in great numbers of single-celled units, i.e. the spores, which are only a few microns in diameter. In *Mucor*, these are formed inside a single

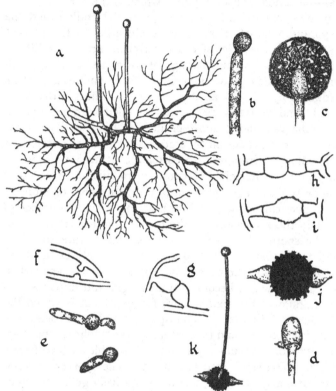

Fig. 18. *Mucor.* *a*, the general habit of the plant shewing the branching mycelium and young erect sporangiophores; *b*, young sporangium; *c*, optical section through an adult sporangium shewing the large columella; *d*, the columella as it remains after the sporangium has burst; *e*, spores germinating by one or two germ tubes; *f—k*, sexual reproduction: *f*, conjugating hyphae originating as opposite bulges on two hyphae; *g*, hyphae in contact; *h*, the gametangia cut off as two cells, one from the apex of each hypha; *i*, the young zygote formed by the fusion of the gametangia; *j*, the mature, thick-walled zygote; *k*, germination of the zygote to give rise directly to a small sporangium. (*f—i*, after Gwynne-Vaughan and Barnes.)

capsule called a "sporangium," which is the terminal compartment cut off from an upright aerial stalk called the
"sporangiophore." Doubtless the elevation of the sporangium above the level of the substratum greatly facilitates
spore dispersal. The sporangiophore arises as an upward-
growing branch of the mycelium. It resembles at first an
ordinary hypha, but as it extends into the air the apex
swells and becomes more densely filled with protoplasm
than the vacuolated non-reproductive hyphae. The dense
apical portion is cut off by a cross wall from the stalk, but
the young sporangium so formed continues to grow in
size, and the nuclei in it continue to divide. Eventually
the protoplasm aggregates into a large number of more or
less spherical masses, each containing a small number of
nuclei; these develop a definite wall and are later set free
as the spores. Meanwhile, as the spores mature, the cross
wall separating the sporangium from the stalk of the sporangiophore swells considerably. Already dome-shaped when
first laid down, it becomes more turgid and expands into
the middle of the sporangium, forming what is called the
"columella" (Fig. 18, c). The ripe sporangium is round and
blackish in colour, and the mature sporangiophores standing
upright from the substratum rather resemble a mass of very
delicate pins stuck into the surface. The sporangiophores
are extremely sensitive, and are negatively geotropic, thus
tending to grow upwards out of the medium containing
the mycelium; they are also positively phototropic in a very
marked degree. It is difficult to see that the latter reaction
has any very definite significance to the fungus, although,
of course, sporangia which project from a dark crevice or
surface towards the light will be most likely to achieve
a free dispersal of spores in the open. As the sporangium
ripens it secretes water, which dissolves the wall and forms
the head into a sporangial drop of liquid, dark with the

ripe spores. From this drop spores are dispersed by splashes of water or by insects: they are not set free to float in the air, although this is the case in some related forms.

When the spores alight on a suitable substratum, they swell up very greatly and germinate at one, two or three points (Fig. 18, *e*). Then from each spore we may see protruding one or more active hyphae which rapidly extend through the medium and give rise to the new fungal mycelium. Within a few hours, under favourable conditions, new sporangia will be in process of formation, and thus reproduction may go on with great rapidity.

(*b*) *Chlamydospores.* In *Mucor*, as in most fungi, under some circumstances spores may be formed somewhat after the manner of the bacterial endospores. The protoplasm of the ordinary vegetative filaments, especially under conditions of water shortage, rounds off into very thick-walled spores, which thus lie inside the old hyphal wall, until it dries and breaks up. These spores are very resistant to adverse external conditions and are to be regarded as resting or perennating bodies, in contrast to the relatively susceptible spores which are produced inside the sporangia, and are essentially agents of multiplication and dispersal.

(*c*) *Conjugation.* The processes of spore formation hitherto mentioned have involved no preliminary nuclear behaviour save ordinary division, and so they may both be termed *asexual* methods of reproduction. In the third type of spore formation in *Mucor* the formation of the *zygospores* is preceded by the fusion of the nuclei in pairs, and this is the essential feature of processes of *sexual reproduction* which are common in all parts of both plant and animal kingdoms, and which will be described in the following chapters. The sexual reproduction of *Mucor* begins with the process of *conjugation*: two large hyphae from the mycelium are seen to approach each other and meet. They both are swollen and

rather densely filled with protoplasm, and shortly the apex
of each is cut off by a transverse wall as in the formation
of a young sporangium. The two cell units so produced are
called the "*gametangia*," for they contain the *gametes* (the
units which fuse together), in the same way as the sporangium contains the spores. Each gametangium contains
many nuclei and when eventually the wall between the two
gametangia breaks down, the nuclei are seen to fuse together
in pairs. These nuclei with their associated cytoplasm are
the gametes. The significance of these terms will become
clearer as the origin of sexual reproduction is more fully
discussed (Chapter IX); for the moment it will suffice to
notice that the gametes are quite similar in appearance
and behaviour so that male and female sexes cannot be
attributed to them. With the fusion of the nuclei they
become enveloped in the very greatly thickened wall of
the single large *zygospore*. This structure might be referred
to by the perfectly general term "*zygote*," which can always
be given to the product of fusion of the gametes in sexual
reproduction. The wall of the zygospore is covered with
rough projections and is so dense that no contents can be
distinguished. On either side of it the remains of the empty
gametangia can be seen. The zygospore, like the chlamydo-
spore, is an extremely resistant resting body. It develops
by the direct production of a sporangium which grows up
into the air and so immediately recommences the asexual
reproduction of the fungus.

A great deal of interest and investigation centres round
the origin of hyphae which will, in the manner indicated
above, fuse to form zygospores. In some species of *Mucor*,
any two adjacent hyphae from the same mycelium can fuse.
In these, zygospores are very abundantly produced, but in
other species so-called + and − strains exist, and hyphae from
one strain will not fuse with hyphae of the same strain but
only with those of the other strain. Here, naturally, zygo-

spores are less frequent. The distinction is probably not comparable with the normal distinction into male and female sexes as found in other plants.

Sporangia and conidia. In the group of fungi which includes *Mucor* a series of forms exists shewing a varying number of spores inside the sporangium. *Mucor* may have several hundred, but others have much smaller numbers, some three or four, and others one only. Where each sporangium contains so few spores, the sporangiophore is often branched, producing many sporangia, and where such numerous sporangia have been reduced to the stage of each containing, or being a single spore, the *conidial* condition is said to have been reached. The single spores thus cut off externally from a hypha are called *conidia*, and the hypha, often branched, is called a *conidiophore*. This type of asexual reproduction is very well illustrated in the two fungi *Peronospora* and *Penicillium*.

Peronospora. *Mucor* could be considered as a type of fungus which is usually saprophytic but which may occasionally become mildly parasitic. In other fungi, such as the bracket fungus, *Polyporus squamosus*, mentioned on p. 95, this capacity for life as a *facultative parasite* is more definitely developed. Some fungi, such as the common mushroom (*Psalliota campestris*) can only exist saprophytically, and these are termed *obligate saprophytes*. Others, of which *Peronospora* is an example, can exist only upon living matter, and are called *obligate parasites*. The rusts of wheat and oats are even better examples of such fungi.

Peronospora belongs to the class of fungi commonly called the downy mildews. They form a fine white web of conidiophores over the surfaces of the plant organs which they infect. Various species of *Peronospora* or closely related genera are responsible for diseases upon an extremely wide range of plants, of which the following are only a selection:

onion, clover, beet and mangold, wallflower, cabbage and turnips, cauliflower, violets, tobacco, maize. Usually each fungal species attacks only one of these plants, or a few related ones. Such fungi are said to shew *selective parasitism*, in contrast to others which are omnivorous.

The damage to the host plant is due to the fact that the greater part of the fungal mycelium is branching through the plant tissues. The hyphae themselves do not directly penetrate the walls of the plant cells, but branch extensively in the long continuous air spaces which separate the cells, and extend especially markedly along the length of plant stems and veins (Fig. 19, *b*). The hyphae in such air spaces send out short lateral branch-hyphae which penetrate the surrounding cells and branch profusely inside them. These then behave as absorbing organs (haustoria). They secrete enzymes which break down the cell protoplasm, and they continually absorb the soluble organic products so produced. The branched haustoria are a characteristic feature of *Peronospora*. The fungus reproduces both sexually and asexually, as is the case with *Mucor*. The asexual reproduction takes place by conidia produced on conidiophores which are branched in a definite bifurcating manner (see Fig. 19, *a*). The conidiophore arises from a hypha which grows out at right angles to the surface of the infected stem of leaf, via the pore of a stoma. As each conidiophore is about 0·2 mm. high the presence of large numbers lends a soft downy appearance to the green surface of the host. The colour is a dingy lilac or white. At the pointed apex of each branch of the conidiophore, a small round swelling is produced, into which nuclei and cytoplasm pass. When the spore has reached its full size and the connection with the conidiophore has been closed the mature conidium jerks off and is blown away. The place where it alights is naturally a matter of chance, but should it fall on the surface of leaves

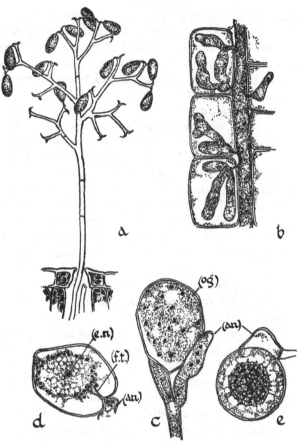

Fig. 19. *Peronospora.* *a*, conidiophore projecting into the air through a stoma of a leaf and bearing conidia. *b*, a hypha running through a vertical inter-cellular space in an infected stem, and putting out thick, branched haustoria into the parenchyma cells on either side. *c*, oogonium (*og*) and antheridium (*an*) each containing numerous nuclei. *d*, the antheridium has put a fertilisation tube (*f.t.*) into the oogonium which at this time contains a single central egg-nucleus (*e.n.*) the other nuclei having taken up peripheral positions. *e*, the fertilised egg has developed into a thick-walled resting zygote. (After Massée, Wager and De Bary.)

or stem of a suitable host plant, when conditions are suitably moist, it germinates at once by the production of a small germ tube which grows over the leaf surface until it enters by turning down into a stomatal pore. From the air-space below the pore, the hypha extends and branches in the air-spaces of the new host.

When much of the host tissue has been killed and the food supplies are diminishing, the fungus begins to reproduce sexually. As in the case of *Mucor*, this involves the fusion of the two gametangia, but here a differentiation of size and behaviour permits a distinction to be made between the *antheridium* or male organ, and the *oogonium*, the female organ. Usually deep in the host tissue the terminal portion of a hypha becomes densely swollen with protoplasm and soon assumes the more or less spherical shape of the typical oogonium. At the same time a lateral branch is formed from the oogonial stalk and grows until it touches the oogonium; this is the young antheridium. Both organs are separated off from the rest of the mycelium by trans-verse walls, and at first both contain many nuclei. Most of the nuclei degenerate, however, so that at maturity there is one male nucleus in the antheridium and one female nucleus (the egg nucleus) in the oogonium. Conjugation takes place by the protrusion of a narrow conjugation tube from the antheridium into the oogonium (Fig. 19, *c* and *d*), and down this some cytoplasm passes together with the male nucleus. The male and female nuclei fuse and the resultant zygote, enclosing most of the cytoplasm of the oogonium, develops an extremely thick and resistant wall. Such zygotes remain in the host tissues until these have rotted away, and they may rest in the soil for several months, to germinate on the ultimate occurrence of suitable conditions.

Pythium. *Pythium* is the generic name under which are

included several species of fungi related to *Peronospora*. One of the best known of them is *Pythium de Baryanum* which is often responsible for the "damping-off" of seedlings over-watered and too closely grown. The plant body consists of a weft of mycelial threads which infest the stem tissues of the seedlings. Unlike those of *Peronospora* these bear no haustoria, but penetrate the parenchyma cells directly and quickly bring about their death and disintegration. As may be imagined the weakened stem of the seedling plant, attacked just above soil-level, quickly collapses and the fungal hyphae branch freely in the dead tissues of the seedling, living saprophytically upon them.

Under especially favourable conditions the fungus may spread from one plant to another by direct growth of the hyphae, but propagation takes place mostly by the formation of conidia or by zoospores. As a hypha reaches the surface of the host-plant it becomes rounded at the apex and densely filled with cytoplasm; it is cut off by a cell-wall and shed as a single spore, to be wind-distributed and to germinate, in suitably moist conditions, by the production of a new mycelium. When the conditions of growth are very moist, or when the fungus is growing on plant material floating in water, the swollen ends of the projecting hyphae behave somewhat differently. Having become densely filled with protoplasm and cut off by a wall from the hypha, they put out a small beak which swells at the end to form a round bladder (vesicle). Into this flow the protoplasmic contents by this time divided up into a number of separate masses. In the vesicle each of these becomes a separate bean-shaped spore, with two laterally inserted cilia (Fig. 20). Since these spores are motile they are known as zoospores. They escape by the rupture of the vesicle, and after swimming about for a time settle down on a new surface, where they germinate directly by formation of a germ-tube. On a suitable host surface infection will take place.

The sexual reproduction does not differ materially from that of *Peronospora*, and the figures of that fungus (Fig. 19, *c, d, e*) will do equally well for both. The sexual organs appear when the fungus is living as a saprophyte on the host it has killed. The zygote is similarly a thick-walled resting body which germinates by forming a new hypha.

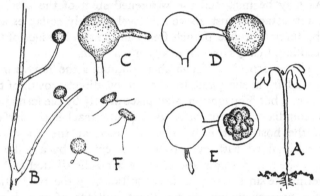

Fig. 20. *Pythium*. *A*, Cress seedling attacked by "damping-off" just above soil-level. *B*, hyphae with sporangia. *C, D*, and *E*, stages in the germination of a sporangium, shewing formation of a lateral vesicle from which zoospores (*F*) are liberated. (After Butler and De Bary.)

It is of interest to compare *Pythium* with *Peronospora* and note the more active parasitism of the former, and the stricter limitation of the latter to land conditions.

Combative methods against fungal diseases. The prevalence of fungi such as *Peronospora* and *Pythium* is increased where great numbers of similar plants are grown together, and since this is the usual practice in modern cultivation methods, measures for controlling fungal disease have come to be very important in agriculture and horticulture. Particularly this is so in tropical countries, where damp and humid climates offer the best possible conditions for fungal growth, with no season, such as the winter in England, during which fungal activity dies down, and after which fresh precautions can be begun. The study of fungal disease has come to be one of the most important aspects of *mycology*, the study of fungi. The measures necessary to check the spread of fungal disease generally depend closely upon the life-cycle of the fungus, as may be seen in the following cases.

(1) *Penicillium italicum*.[1] This fungus is responsible for the blue mould which is so commonly seen on oranges, lemons and grape fruit; the rot which it produces on these citrus fruits is of the greatest commercial importance and has evoked various protective measures. These depend in part upon the fact that the mould is a *wound parasite*. Once in contact with the living cells of the orange fruit it will spread swiftly. It cannot of itself penetrate the thick oily skin, but it relies for entry on accidental openings. This explains why oranges may lie in a box embedded in a rotten blue-green mass of decayed fruit, and yet be quite sound. It also indicates that the disease may be limited by care in gathering and packing so that bruising and cutting of the skin may not occur. Almost all citrus fruits are washed before packing, and it is now a common custom to add to the washing water some substance such as borax, which, on drying, remains in the crevices of the skin and will destroy fungal spores in those places even if slight wounds should be present, or should develop later.

---

[1] *Penicillium* is a fungus of the most widespread occurrence on all kinds of organic material, jam, leather, fruit, etc., and it gives rise to bluish-green moulds, of the colour familiarly seen in Stilton cheese, where it is produced by the growth of certain species of this genus. This fungus reproduces asexually by conidia borne in long feathery chains at the end of conidiophores, which are branched in a finger-like way. These conidiophores have a very characteristic appearance (Fig. 19, *a*). *Penicillium notatum* is the fungus from which the substance penicillin is produced (see p. 140).

(2) *Phytophthora infestans* (Fig. 21, *b*). This fungus is
responsible for the potato blight which caused the great
Irish famines about the middle of the last century, and
which often still causes considerable damage to English
potato crops.

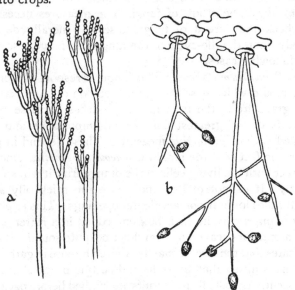

Fig. 21. *a*, conidiophores of *Penicillium*; the chains of spores (conidia) are
often much longer than this but they very readily break with any manipulation.
b, conidiophores of *Phytophthora infestans* (potato blight fungus), projecting
through stomata of the lower epidermis of a potato leaf and producing conidia.

The precautions against its spread reflect very closely its
life conditions. The fungus perennates as a mycelium in
the young potato tubers, and grows up in the young shoots
in spring. From this shoot conidia are distributed.[1] They

[1] The reproduction of *Phytophthora* is rather more complicated than is
indicated here. Full account of the disease and methods of dealing with it are
given in *Plant Diseases*, by F. T. Brooks, Oxford University Press.

rapidly infect leaves and stems, and the fungus mycelium produces a swift black rot of the haulm, which may end in a severe loss of crop. The conidia also infect the young tubers in the soil, and it is therefore advisable to use only "seed" tubers from healthy crops. As the crop develops it is advisable to spray the haulm with "Bordeaux mixture," a suspension of basic copper salts, made by mixing milk of lime and copper sulphate solution. This forms a thin film on leaf and stem surfaces, which kills the conidia as they germinate. Different varieties of potato vary somewhat in their susceptibility to the disease, but no completely immune varieties are known. In the case of wart-disease of potatoes, which is due to an entirely different fungus, it is possible to avoid the disease completely, even in infected soil, by growing immune varieties, and in fact there is legislation in this country which forbids growing non-immune varieties in infected soil. By plant-breeding and systematic trials many immune varieties have been made available, such as Witch Hill, Immune Ashleaf, Arran Comrade, Arran Crest, Herald, Kerr's Pink and Majestic. Among the older susceptible varieties are Sharpe's Express, King Edward and Up-to-Date.

Fungi parasitic on animals. It is generally true that the organisms responsible for disease are bacteria in the case of animals and fungi in the case of the green plant, but a few exceptions occur in both cases.

Saprolegnia. Dead organic material which occurs in ponds and streams, such as dead fish or flies, is often invaded by the fungus *Saprolegnia*, as well as by myriads of bacteria. The fungus can also behave as a facultative parasite and as such it attacks carp and gold-fish, sometimes where scales have been broken off, and sometimes at the gills. It blocks up the gills, and the fish may be killed in this way. The fungus reproduces asexually by means of sporangia which

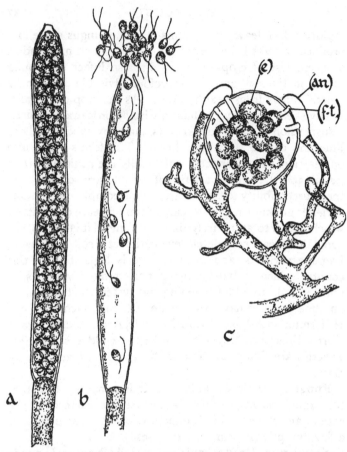

Fig. 22. *Saprolegnia*, a fungus parasitic and saprophytic on animals. *a*, the terminal cell of a hypha cut off by a cross-wall to form a sporangium. The protoplasm inside it has rounded off into spores. *b*, the mature sporangium by opening at the apex has liberated the spores into the surrounding water. The spores are bi-ciliate and motile. *c*, the sexual reproductive apparatus shewing a single spherical oogonium containing mature eggs (*e*), and closely enveloped by several male organs, antheridia (*an*). The antheridia have sent fertilisation tubes (*f.t.*) into the oogonium and through them the antheridial contents (male nuclei) have passed to the eggs and fertilised them. (*a* and *b*, after Thuret; *c*, after Pringsheim.)

grow out into the water (Fig. 22, *a* and *b*). They appear as stout club-shaped hyphae, inside the ends of which the protoplasm divides up to form spores. These are colourless ovoid bodies which are motile, possessing two delicate cilia, and are set free directly into the water by the apical bursting of the sporangium. Sexual reproduction resembles that of *Peronospora* fairly closely, save that several egg cells are produced in each oogonium, and the antheridium is a branched structure which encircles the oogonium, putting separate fertilisation tubes into the several eggs (Fig. 22, *c*). Also in some species the egg cells may develop without any previous fertilisation, a proceeding known as *parthenogenesis.*

Human diseases. The commonest fungal diseases from which man suffers are skin diseases, some of them of such slight importance that they are almost disregarded. Such is the condition of *Pityriasis versicolor,* an infection, mostly saprophytic, by the fungus *Microsporon furfur,* of the skin surface. Another species of *Microsporon* is responsible for most cases of ringworm of the scalp in children. The fine mycelium occurs in breaks in the dermis; it infects hairs where they emerge from the sheath, and grows up and down them. The hyphae in the hair put out to the surface fine filaments on which spores are borne. These are very small (2μ) and are produced in great numbers. They are readily detached, and this contributes to the infectiveness of the disease. *Microsporon andouïni* is the commonest species, but allied forms are found on cats, dogs, horses, etc., and these may also infect man. Various species of *Trichophyton* are responsible for other forms of ringworm usually more severe, though of the same general type as those due to *Microsporon.* The rather similar affections of skin and hairs classed as favus are attributed to species of *Achorion.* The diseases known as thrush, aspergillosis, sporotrichosis and blastomycosis are all due to fungal infection of greater or less severity. Possibly the most serious fungal disease is that called "Madura foot" which is prevalent in India and other tropical countries. It principally affects the foot, and causes ulceration, granulation and necrosis of the bones, so that the disease becomes extremely chronic and incurable save by operation.

Penicillin. One of the great medical discoveries of recent times is that of the production by the fungus *Penicillium notatum* of a substance, known as penicillin, which has astonishing powers of preventing the growth of bacteria, among them types responsible for many serious human diseases. The discovery began with the observation that in a zone round a single colony of *Penicillium notatum,* present accidentally on a plate of bacterial cultures, all bacterial growth was prevented. This was followed at a later stage by the purification of the substance penicillin from the culture medium, and the very important discovery that it was non-toxic to human beings. An enormous industry has now been developed for the cultivation of *P. notatum* in vast amounts and free from contamination by other micro-organisms, and for adequate extraction and purification of the penicillin produced by it.

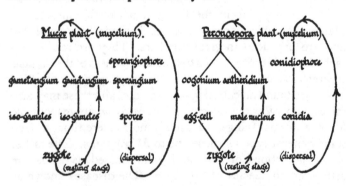

Diagram of life-cycle of *Mucor* and *Peronospora.*

CHAPTER IX

THE GREEN ALGAE

The fungi, although shewing all degrees of variation in organisation between the single cell of yeast, the mycelium of *Mucor* and the bulky and complex mushrooms and toadstools, never achieve a differentiation of the plant body into organs comparable with the root, stem and leaves of the higher plants. For this reason they form part of the large plant group, the *Thallophyta*, in which the plant body may always be referred to as the *thallus*. The bacteria and the algae also belong to this group, whilst the ferns and all the seed-bearing plants, including both coniferous tree types and flowering plants, belong to the more complex group, the *Cormophyta*. The liverworts and the mosses are placed in a group shewing intermediate complexity, called *Bryophyta*.

The algae are distinct from the fungi in that they possess chlorophyll, and so are able to live holophytically. They are also an essentially aquatic group of organisms, and they include tiny unicellular, free-swimming plants like *Chlamydomonas*, as well as the large red, brown and green seaweeds common within the tidal zone of the seashore. Even in these coloured forms, chlorophyll is present, although masked by the presence of other pigments. The range of structure is considerably greater than that found in the fungi. They occur very widely in both fresh and salt water and even occur to a limited extent on land, playing some part in the organic life of the soil. Great interest is given to the algae from the fact that they include many types similar to those from which in past evolutionary history all the present green plants may have developed. Equally with animal life it is thought probable that the origin of plant life was in the sea. Some small protoplasmic unit, like a modern flagellate, may have commenced by the acquisition of chlorophyll, the evolution of the whole of the green plant world, giving rise in succeeding ages to forms of increasing complexity of organisation, and of diminishing dependence upon the water which entirely surrounded their flagellate ancestor. In the end, green flowering plants have been produced which can live and reproduce in extremely arid situations, although in all of them some access of water to the roots is essential, for still all the cells which constitute their tissue can only live when supplied with water. An alga which may be regarded as like the primitive flagellate is *Euglena*, which has already been described, but it is more usual to trace plant ancestry to a unicellular free-swimming form with a definite cellulose wall such as *Chlamydomonas*, for the cellulose skeleton is a character which is practically constant throughout the plant kingdom.

In the following pages we have to consider a number of simple types of present-day green algae, which shew, when arranged suitably, a range of increasing complexity and development in several characters, which may be taken to represent tendencies of evolution in the green algae, both of the present and of past times. In particular they shew the origin of the *soma*, the *body* of the organism as distinct from its reproductive cells, and the origin of differentiation of conjugating cells into the two sexes.

Chlamydomonas. The structure and appearance of *Chlamydomonas* have been already described on p. 11. The adult organism is a single ovoid cell with cellulose wall, cilia, chloroplast, nucleus, pyrenoid, eye-spot, contractile vacuole, etc., and it is physiologically complete and capable of independent existence.

Vegetative reproduction. As the cell increases in bulk by the assimilation of inorganic salts and the carbohydrates formed in photosynthesis, a limiting size appears to be

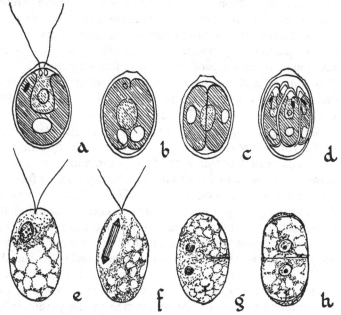

Fig. 23. Vegetative division. *a—d, Chlamydomonas.* The chloroplast is indicated by continuous shading and the central cytoplasm as seen through the chloroplast by broken shading; the pyrenoid is left white. In *d*, four daughter cells have been formed inside the parent cell. The nuclear behaviour is obscured in *Chlamydomonas* by the chloroplast. In *Polytoma* (*e—h*) there is no chloroplast and nuclear division and cell-wall formation can be readily observed. *f*, shews spindle fibres in the dividing nucleus. (Figures after Oltmanns.)

reached, at which division into two or more new individuals of smaller size takes place. These in turn grow to the limiting size before dividing. In the process of division, the cilia are withdrawn, the cell comes to rest, and the protoplasm contracts into the middle of the cell. First the nucleus, and then all the cytoplasm, divides by a furrow growing inwards from the cell-wall and separating the cell into two equal daughter cells, each of which is a new daughter individual with chloroplast, cilia and cell-wall like the parent cell. These daughter cells begin to move within the cell-wall of the parent and eventually burst out from it and begin an independent existence. In some cases the process of division involves a second bipartition following the first, so that four daughter cells are produced within the parent cell. These are naturally smaller than the daughter cells produced two at a time (Fig. 23).

Sexual reproduction. The division we have considered is a method of *vegetative* reproduction, since it occurs by a simple continued growth of the body of the plant, and it is *asexual* because not associated with any preliminary nuclear fusion. In some circumstances, however, *Chlamydomonas* reproduces sexually. A process of bipartition takes place like that in vegetative reproduction, but it is many times repeated, so that in each parent cell a number (varying from eight to sixty-four) of small replicas of the parent cell is produced. These, in some species, are naked protoplasm and in some have a cellulose wall, and they vary in size according to the numbers in which they are formed in the parent cell. These small cells are the gametes. They are freely motile like the parents, and they conjugate together in pairs; two gametes approach and fuse together at their clear ciliated ends to form one round body with four cilia. This may swim about for a time, but soon the cilia are withdrawn, and a spherical thick-walled resting body is produced, which can withstand considerable drying up like the zygotes of *Mucor* and *Peronospora*. On germination in suitably wet conditions the zygote divides vegetatively into two or four daughter cells.

Sphaerella (*Haematococcus*). An organism much resembling *Chlamydomonas*, and often easier to procure is *Sphaerella*, another unicellular green alga common in ditches, pools and rain-water tubs. The cells shew the same basin-shaped chloroplast as *Chlamydomonas* (though more broken up and irregular), with the single red eye-spot, and two long equal cilia at the apex. The cytoplasm however contains rather abundant contractile vacuoles, and with the chlorophyll there is often found a red pigment, haematochrome, some-times so abundantly that the whole cell is deep red in colour. It is the wall-structure however which is of particular interest, for the cellulose in adult cells swells into a wide mucilage-layer almost as thick as the green centre of the cell itself. Through this wide wall project both the apical cilia, and fine strands of protoplasm also radiate from the cytoplasm most of the way through it, sometimes dividing at their extremities (Fig. 24).

In reproduction *Sphaerella* does not differ essentially from the *Chlamydomonas* type. It shews an asexual re-production in which daughter cells form by division of the parent cell into two or four similar individuals. In sexual reproduction (not observed in the species found in Britain) similar small motile cells fuse together in pairs and form zygospores. As many as a hundred such gametes may be formed in one parent cell. Very often it is possible also to find resting stages of this alga; the protoplasm rounds off, the cilia withdraw and a firm external layer forms. As this happens haematochrome accumulates, so that the resting cells are a deep rusty red colour, and if present in great numbers they lend the water a marked reddish tinge. If the dried mud from a pond containing *Sphaerella* is caught up as dust by heavy winds it may cause the phenomena known as "red rain," or the alpine "red snow." This fact demonstrates at once the importance to the green alga inhabiting shallow fresh water, of a stage resistant to drought, in which both perennation and dispersal may be brought about. The marine algae such as *Fucus* have no such stages.

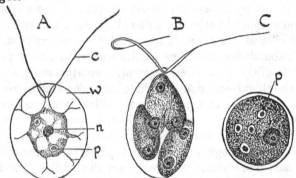

Fig. 24. *Sphaerella* (Haematococcus). *A*, motile cell with cilia (*c*), gelatinous wall (*w*) penetrated by strands of protoplasm, central nucleus (*n*) and pyrenoids (*p*). *B*, cell still motile containing four daughter cells. *C*, thick-walled dark-red resting cell. (After Fritsch and Salisbury, Wollenweber.)

Differentiation of sex. In many species of *Chlamydomonas* the gametes produced are all of the same size and they are called *isogametes*, and their conjugation is called *isogamy*, but in other species varying numbers of divisions in the parent cells give rise to gametes of all sizes (*heterogametes*). In such a case conjugation takes place only between a large and a small gamete, a process called *heterogamy*, since the fusing gametes are unlike (Fig. 25). In this case the difference is one of *size* only, but we shall find in later examples that large size tends to go with passivity and small size with greater motility, so that to size we must add a difference in *activity* also. To these we shall also find added differences in *structure*, but for this purpose we must examine the members of a series in which the algal cells produced by division do not separate as in *Chlamydomonas*, but remain aggregated together in a colony or *coenobium*.

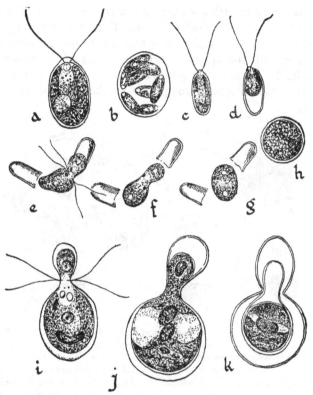

Fig. 25. Stages in the conjugation of *Chlamydomonas*. *a—h*, *C. media.*
a, vegetative cell; *b*, eight gametes produced inside the parent cell; *c*, single
gamete; *d*, gamete before conjugation shewing protoplast at one end of the
cell; *e, f, g*, conjugation between two gametes of unequal size. The protoplasts slip out of the cellulose cell-wall at the time of fusion, the cilia withdraw and there is a complete fusion of the protoplasts. *h*, the resultant thick-
walled zygote. (After Klebs.) *i, j, k, C. Braunii* (after Goroschankin). In
this species the gametes fuse whilst yet within their cellulose walls. There is
marked heterogamy; the protoplast of the small gamete slips into the cell of
the larger gamete. *j*, shews the nuclei, chloroplasts and pyrenoids of the two
gametes still separate; *k*, shews the fused nuclei.

Pandorina is such a colony of sixteen cells, each like a single *Chlamydomonas* cell (Fig. 26). The cells are arranged in the form of a solid sphere and are rather pyramidal in shape, with the apex towards the centre of the sphere. The flat basal outer part of the cell shews an eye-spot, and the two cilia project through the mucilaginous envelope

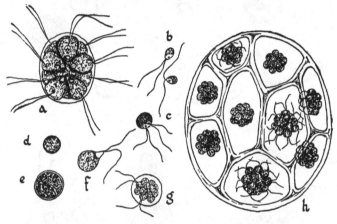

Fig. 26. *Pandorina.* *a*, the adult colony of sixteen similar ciliated cells; *h*, a coenobium undergoing asexual reproduction—each cell has divided to form a daughter coenobium which still remains within the parent body. Some of the colonies have already produced cilia, and will shortly break out of the parent cell. *b*—*g*, stages in sexual reproduction—*b*, motile gametes; *c*, stage immediately after fusion of two gametes; *d*, later stage shewing cilia withdrawn; *e*, later stage shewing resting zygote with thickened wall; *f*, motile cell produced by the zygote on germination; *g*, new colony produced by vegetative division of the motile cell.

which binds the whole colony together. In asexual reproduction each cell of the colony simultaneously begins to divide, and by four rapidly following bipartitions each cell produces within itself sixteen smaller cells, which, enclosed in a common mucilage, constitute a daughter colony, and escape as such by the breaking up of the parent colony. It is interesting to note that all the sixteen cells of the parent colony produce daughter colonies in this way. The same thing holds for the sexual reproduction, in which all the cells participate by the formation of gametes. By successive divisions gametes of two sizes are produced, the large ones sixteen, and the small ones thirty-two per cell, and these fuse without reference to size, so that sometimes the fusing gametes are similar (isogamy), and sometimes dissimilar (heterogamy).

Origin of the soma. We have commented already on the fact that each cell of *Pandorina* is capable of reproducing sexually or asexually in exactly the same way as the individual *Chlamydomonas* cells. Each vegetative cell resembles every other cell, and is capable alike of all the normal cell metabolism, and of reproduction after reaching a certain size. Thus the cells merely multiply indefinitely, and none of them need die, so long as conditions remain favourable. They do not die "of old age" but remain potentially immortal. Though this condition is prevalent among unicellular organisms, such as yeasts, bacteria and flagellata, both animal and plant in nature, it by no means holds in the higher plants and animals. In these there is a definite "mortal body," or "soma," which dies of senescence, however favourable conditions may be, and the only parts which persist are the reproductive cells (or germ cells) which give rise to new organisms of the same kind. It is possible to exemplify in the algal series we are considering how such a soma may have originated, and indeed how mortality itself may have originated. Two forms of coenobiate algae, *Eudorina* and *Pleodorina*, illustrate the transition between *Chlamydomonas* and *Pandorina*, which have no soma, and *Volvox*, which, as we shall see later, has a very well-developed soma.

Eudorina is a spherical colony containing thirty-two *Chlamydomonas*-like cells, arranged in the form of a hollow sphere with a wall one cell thick. As in all these spherical colonial types, the colony begins development as a circular plate, which curves round till it unites into a sphere, leaving a small pore where the closing up is not quite complete. The cells are capable of vegetative division to form daughter coenobia, or gametes.

Pleodorina is a globose or ellipsoidal colony extremely like *Eudorina* in general appearance, and containing 32, 64, or 128

somatic cells

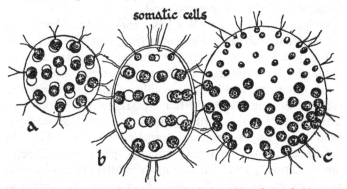

Fig. 27. The Soma. *a, Eudorina*, a spherical motile colony of thirty-two similar cells all capable of division; *b, Pleodorina illinoiensis*, a spherical motile colony consisting of thirty-two cells, of which four at one end of the colony constitute a "soma," which dies when the other twenty-eight cells divide; *c, Pleodorina californica*. The somatic cells constitute about half the colony. (After West and Fritsch.)

cells. As in *Eudorina*, the unity of the colony is emphasised by the fact that the cells apparently so loosely embedded in mucilage are connected together by delicate protoplasmic threads. There are two species of *Pleodorina*, and in both a definite soma is present (Fig. 27). In *P. illinoiensis* the coenobium contains thirty-two cells, of which twenty-eight are exactly like those of *Eudorina*, whilst four smaller cells, grouped together at one end of the colony, form the soma, for they are incapable of dividing, and die when the rest of the organism forms daughter colonies or gametes. *Pleodorina*

californica consists of 64 or 128 cells of which from one-third
to one-half are somatic, and as these somatic cells are grouped
together, they give the colony a characteristic appearance
(Fig. 27). The last member of this series of motile coenobia
is *Volvox*, which is made up of a very large number of
Chlamydomonadine cells (up to 25,000), arranged in a
hollow sphere one cell thick. The English species grow
almost to 1 mm. in diameter, and we may imagine that
they represent the mechanical limit of size for organisms
constructed on these lines. Of the very large number of their
component cells, only a small percentage, usually less than
1 per cent., are capable of dividing to form either daughter
colonies or gametes, and the rest of the cells form the bulk
of the plant body, as is the case with all the higher plants
and animals. Somatic and reproductive regions are not set
apart as in *Pleodorina*, but the reproductive cells occur
scattered separately through the soma though most abundantly in one-half of the colony (see Fig. 28). It is not to
be thought that because the *Chlamydomonas—Pandorina
—Pleodorina—Volvox* series illustrates the evolution of the
soma so well that these forms are on the direct line of
evolution of the higher plants: a similar evolution of the
soma must have occurred many times along various evolutionary lines, both of plants and animals. In the higher
organisms it is easy to see how far the origin of the soma
is a differentiation of cells with vegetative functions from
those with reproductive functions: a definite division of
labour. The distinction has become particularly complete
in the higher animal, in which the reproductive cells are
set aside at an early stage in development, and the somatic
cells have lost all powers of reproducing the species. This
is not quite so marked in the higher plants, for not only can
new reproductive cells arise from time to time in the life of
a single plant, and in various parts of it, but the somatic

vegetative cells have not lost all powers of reproduction, and a small mass of cells from the stem, or even from the leaf or root of a flowering plant, will often give rise to a new plant, by so-called "vegetative" growth. This difference between plant and animal cells is undoubtedly related to the fact that in general the former are far less closely specialised to particular functions than the latter.

Volvox. *Structure.* The structure of the volvox sphere can probably best be made out by reference to the figures on pp. 155 and 157. There are two species of *Volvox* commonly found in Britain, and they differ slightly in structure. They are *Volvox aureus* and *Volvox globator.* When living, *V. aureus* looks like a large sphere of clear jelly, in the surface of which green dots are uniformly scattered. By staining the colony with methylene blue or gentian violet the reason for this appearance can be made out. Each green dot will be recognised as the protoplast (i.e. the living protoplasm, chloroplast, nucleus, etc.) of a single cell, and it will be seen to be contained in a polygonal area, outlined by a thin deeply stained wall. This is the primary cell-wall enclosing each of the cells of the colony, and the clear space between it and the protoplast is a mass of mucilage produced in the further development of the cell-wall. The staining will probably reveal fine protoplasmic threads traversing the mucilaginous and primary walls, from each protoplast to its neighbours, putting the whole into protoplasmic continuity (Figs. 29 and 30). *V. globator* has a similar structure, but the protoplasts appear star-shaped rather than round, for in place of thin, protoplasmic filaments running between them, stout projections of protoplasm reach from one to the next (Fig. 29). Both species shew a similar general structure, so that a section through the sphere would shew a single firm outer layer of cells with cilia projecting outwards, and the space inside

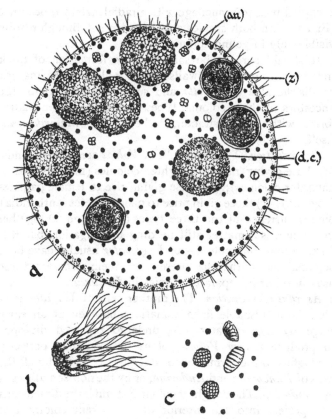

Fig. 28. *Volvox aureus. a,* a medium-sized colony shewing as round black dots the numerous somatic cells of which it is made up; the protoplasmic connections between them and the cell-walls can only be made visible by staining. The colony contains three types of reproductive units; daughter colonies (*d.c.*) produced asexually by division of a single cell; ripe egg-cells or young zygotes (*z*); and young antheridia (*an*) whose contents are dividing up and will eventually form sperms. *b,* a colony of ripe sperms which has just escaped from the antheridium; *c,* mature antheridia as seen in surface view of a colony; in two the sperms are seen sideways and in two, endways. (After Klein.)

occupied by a soft mucilage with a radial striate appearance (Fig. 30). In both species a pore is present, though readily visible only in young colonies.

It should perhaps be said that the formation of thick mucilage from the cell-walls as the cell becomes mature, recalls the alga *Sphaerella*, and suggests that perhaps the ancestors of these colonial types were Chlamydomonadine forms with this character, rather than *Chlamydomonas* itself.

Reproduction. The germ cells of *Volvox* are of three kinds: (*a*) those giving rise to daughter colonies (mother cells of daughter colonies); (*b*) those giving rise to female gametes (oogonia); (*c*) those giving rise to male gametes (antheridia). The occurrence of the three types (*a*), (*b*) and (*c*) together in colonies is a matter still needing investigation, but it is fairly certain that colonies of *V. globator* usually have both oogonia and antheridia present in them, whereas these usually occur in separate colonies in *V. aureus*.

Asexual reproduction. The mother cells of daughter colonies, distinguishable from somatic cells even at an early stage by their greater size, undergo repeated divisions to produce a new *Volvox* colony, in a manner perfectly analogous to the formation of daughter colonies by all the cells of *Pandorina* and *Eudorina*, or by the non-somatic cells of *Pleodorina*. The daughter colonies attain large dimensions and project into the interior of the parent colony, and becoming detached, they may swim about inside it until they are set free when it dies and breaks up (Figs. 28 and 30). This may be delayed so long that the daughter colonies may then themselves contain daughter colonies, and even more generations than three may be found inside large and old *Volvox* colonies. It is worth while noting that the young colonies are quite dark green in colour, for the walls are still thin. As they age the walls grow mucilaginous, and their

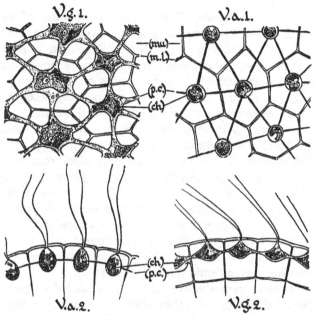

Fig. 29. *Volvox.* Diagrams to shew the structure of the colony of two species of *Volvox.* *V.a.*1., surface view of a small part of the colony of *V. aureus*; *V.a.*2., section through a similar region; *V.g.*1. and *V.g.*2. shew *V. globator* in the same way. The protoplasts are very different in shape in the two species but in both they have been separated by the formation of mucilage (*mu*) by the cell-walls; the unaltered middle layer of the walls (*m.l.*) is still visible. Protoplasmic strands (*p.c.*) fine in one species and thick in the other connect the protoplasts. Otherwise each protoplast with its single curved chloroplast (*ch*), eye-spot and two cilia has the structure of a motile unicellular alga such as *Chlamydomonas.* (After Janet.)

increasing width carries apart the green cells, so that the colonies look paler.

Sexual reproduction. The sexual reproduction of some species of *Chlamydomonas* and of *Pandorina* has been described as shewing some slight differentiation in size between the sexual gametes; here in *Volvox* a much greater separation is evident. In both species of *Volvox*, oogonia occur among the somatic cells as larger non-ciliate cells with an intense green colour, and a thick mucilaginous membrane. They project somewhat into the interior of the colony (Fig. 30, *e*). Each oogonium produces only a single gamete which is large and inactive, and stored with food reserves and contains chlorophyll. All these characters stamp the gamete as female, and because it is non-motile it is referred to as an egg cell. It is fertilised in the cell in which it is produced by the conjugation with it of the male gamete, and prior to this it derives a great deal of nourishment from the cells of the parent colony, with which it remains in contact by numerous protoplasmic threads (Fig. 30, *b*).

The male reproductive cells, the antheridia, undergo repeated divisions, until thirty-two cells have been formed inside the parent in the case of *V. aureus*, and 128 in *V. globator*. These are the male gametes, and they differ very considerably from the female gametes. They are very much smaller even than the ordinary somatic cells, and are small cigar-shaped bodies, lying together parallel in a bundle in *V. aureus* (Fig. 28), and arranged in a small sphere in *V. globator* (Fig. 30, *e*). In *V. aureus* they are green and have two cilia attached to the end of the tapering body, but in *V. globator* the chloroplast has become a small yellow plate no longer functional, and the cilia are attached to the middle of the body. The plate or the sphere of male gametes (or sperms) is set free in the interior of the parent colony, where it swims about in the thin gelatinous

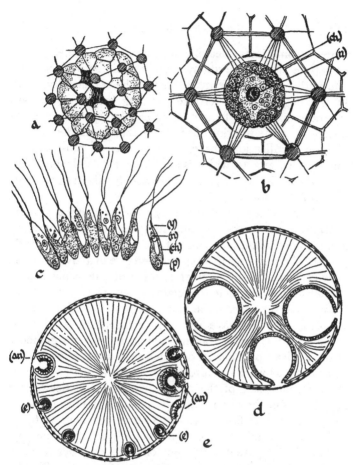

Fig. 30. *Volvox. a, V. aureus,* a daughter colony of small size seen through the layer of cells of the parent colony; the opening of the young colony is shaded. *b, V. aureus,* a single egg cell among the ordinary somatic cells; abundant protoplasmic filaments connect it with surrounding cells and it contains large nucleus (*a*) and chloroplast (*ch*). *c, V. aureus,* a plate of mature sperms just liberated from an antheridium and now beginning to separate. Each contains nucleus (*n*), eye-spot (*y*), cilia, a chloroplast (*ch*), and pyrenoid (*p*). *d, V. globator,* diagrammatic section through the middle of an old colony shewing three large daughter colonies (*d.c.*) projecting into the interior of the parent colony which is full of thin mucilage with a radiating structure. *e, V. globator,* similar section to *d,* shewing three antheridia (*an*) in different stages of maturity and three large egg cells (*e*). Both types of organ have been formed from a single cell of the parent sphere into the interior of which they now project. In *d* and *e,* the cilia of the somatic cells have been omitted. (After Janet and Klein.)

matrix, sometimes breaking up into separate sperms and sometimes not. If the colony is one containing both oogonia and antheridia, the sperms, attracted chemotactically, swim to the oogonia, and fertilise the egg cell. If there are separate male and female colonies, the sperms have to escape from the interior of the male colony via the pore.

After fertilisation, the egg forms a dark, red-coloured, thick-walled zygote, which is only liberated by the breaking up of the parent colony. It germinates by direct division to form a new colony.

Differentiation of sex (contd.). *Volvox* shews a greater differentiation in structure and function, between the male and female gametes, than any of the types so far considered. The small differences in size, and therefore in activity, which have occurred in the simpler types have given place to a specialisation in which the male gametes are small, very active cells, produced in large numbers, and capable of only a short life, for they have little or no chlorophyll. Their only function seems to be transmission to the female gametes of the nuclear protoplasm of which they so largely consist, and which, as we have already seen, is the material basis of the transmission of the hereditary characteristics of the organism. On the other hand, the female gametes are large and passive, and few in numbers; they are densely filled with reserves, and often contain chlorophyll. By remaining *in situ* in the parent body, not only do they derive food materials from it before fertilisation, but after it also. Should germination take place *in situ*, the embryo organism also derives nourishment and shelter from the parent colony. The structure of the egg cell is thus seen to be closely connected with its particular function of providing the young organism with adequate nourishment and protection. With the passivity and fertilisation *in situ* of the female gamete, there has developed a co-ordination between the male and female gametes; the male has become sensitive to very small amounts of chemical substances secreted by the female cell. The male gamete makes the chemotactic response of swimming towards the higher concentrations of these substances, and so is enabled to "seek out" the egg cell and fertilise it.

All this complication of specialisation into male and female gametes, which is so evident in the higher animals and plants, we have here visualised as evolved within the one small algal series we have described. It is not to be thought that in actual fact *Chlamydomonas* gave rise to *Pandorina*, *Pandorina* to *Eudorina*, *Eudorina* to *Pleodorina*, and *Pleodorina* to *Volvox*. All these are present-day forms and we do not know their ancestors. All that we can say is that they appear to be at different stages of evolution in respect of certain characters such as size of colony, differentiation of the soma, and differentiation of the sexes, and by considering them as a sequence, we may be helped in our suppositions as to the manner in which evolution from unicellular organisms may have taken place.

Spirogyra. We have already indicated that primitive unicellular plants shew a varying tendency to separate or aggregate when in a dividing condition. The latter tendency must have been a factor in the evolution of coenobiate forms of the *Volvox* series, but other arrangements of the cells were possible besides the sphere, and so plates and filaments and irregular gelatinous masses of cells also came into being, some of them motile, but the majority of the typical non-motile plant type. *Spirogyra* is such a non-motile filamentous type. It consists of a simple unbranched thread made up of cells all exactly alike in appearance, usually longer than broad, and joined end to end. It is a particularly common alga, forming the delicate threads which lie on the surface of ponds and ditches, in bright green masses, often several feet across. It is identifiable by the sliminess of the threads to the touch, a character due to the mucilaginous nature of the outer layers of the cell-wall. Each cell is cylindrical in shape, and the cellulose wall is lined with a layer of cytoplasm which encloses the large central vacuole. Embedded in the cytoplasm is a spiral band-shaped chloro-

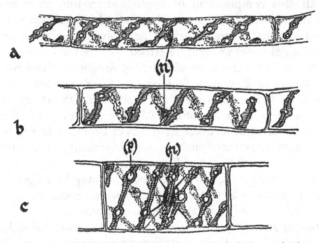

Fig. 31. *Spirogyra*. Three species are shewn, all of which contain the spiral chloroplast (shaded), which is characteristic of this alga. The pyrenoids (*p*) are shewn unshaded. Suspended in the middle of each cell by protoplasmic strands is a definite nucleus (*n*). *a*, species with two chloroplasts; *b*, species with one; and *c*, species with three.

plast winding round the cell from end to end. At a casual glance the cell seems to have a crossed lattice-like arrangement of chloroplasts inside it, but by careful focussing this can be readily resolved into the spiral form. According to the species of *Spirogyra*, the chloroplast may be drawn out or compressed, and in some species there are several parallel chloroplasts in each cell (Fig. 31). The margin of the chloroplasts is usually irregular, and borne on the chloro-

plast, often regularly along the middle of it, is a series of
very brightly refractive pyrenoids. By staining in iodine
threads which have lately been illuminated, a ring of starch
can be made out round each pyrenoid. This suggests that
these bodies may play some definite part in starch formation. Starch does, however, occur elsewhere in the chloroplast. The nucleus of the cell is suspended in the middle of
the vacuole by numerous threads of protoplasm, often
spoken of as "bridles," which connect up with the cytoplasm
and very often join on to the chloroplast at the places where
the pyrenoids occur. The nucleus is clearly visible in living
cells where it happens not to be hidden in the turns of the
spiral chloroplast.

Reproduction—vegetative. The filament is able to increase
in length by the division of all its constituent cells, and as
the filaments very readily break up, "fragmentation" can be
regarded as a very effective means of vegetative reproduction of the alga. In division the nucleus divides by mitosis,
the two nuclei go into opposite ends of the cell, and
a wall begins to form between them, growing inwards like
a closing iris diaphragm and eventually cutting the cell in
two in the plane transverse to the length of the thread. The
new wall cuts the cytoplasm and chloroplasts directly, so
that the latter seem to run continuously in their spiral from
cell to cell (Fig. 32).

Sexual reproduction. As in the case of *Mucor* and
Peronospora in the fungi, *Spirogyra* produces gametes
which are non-motile, within gametangia which themselves
fuse with each other. In the process of conjugation two
filaments of cells come to lie parallel alongside each other,
and very often every cell of the two filaments becomes a
gametangium. The cell-walls of two cells lying opposite to
each other in the two filaments become attached locally and
begin to swell at these points, producing a short tube which

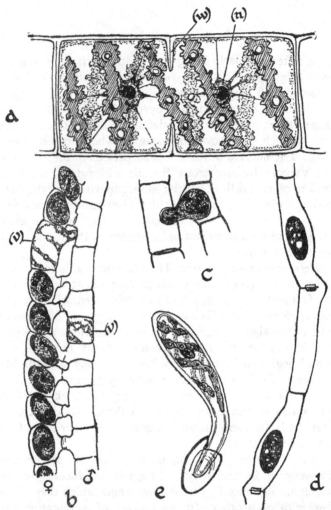

Fig. 32. *Spirogyra. a*, asexual reproduction (simple cell-division, which may occur in any cell of the filament). In the cell shewn, division is almost complete, two daughter nuclei (*n*) are present, and a new cell-wall (*w*) will shortly cut across the chloroplast and separate the two daughter cells by a continuous plate. *b—e*, sexual reproduction. *b*, "ladder"-conjugation. The cells of two filaments lying side by side have conjugated, those in the right-hand filament (marked ♂) behaving as male cells, and those of the left-hand filament as female (♀). One cell in each filament has remained vegetative (*v*) there being no corresponding gamete available in the opposite filament. *c*, passage of the gamete through the conjugation tube. *d*, "chain"-conjugation between neighbouring cells of the same filament. *e*, germination of zygote. (*d* and *e* after Tröndle.)

forces the filaments apart (Fig. 32, *b*). Finally, by breakdown of the transverse wall across the tube, connection between the two cells is established. Meanwhile each cell (or gametangium) has formed a single gamete inside itself by the rounding off of all the cell contents. The gamete of one cell then flows through the connecting tube into the other cell, and there the two gametes fuse, giving rise to a thick-

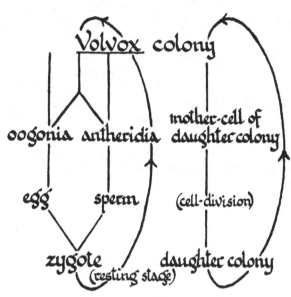

Diagram of life-cycle of *Volvox*.

walled, ovoid zygote. The only distinction between the two gametes lies in their activity, the non-motile one being regarded as female. It usually happens that when two filaments are conjugating, one will be entirely made up of female cells and the other of male cells, so that at the end of the process the one will contain a row of zygotes and the other will be empty. Thus it is possible to speak of male and female filaments, but the distinction is quite slight, and breaks down in many cases. Thus a filament may often be found, male in one part and female in another, zygotes may be formed in the conjugation tube (so indicating isogamy), and in some species "chain"-conjugation occurs in place of the usual "ladder"-conjugation, conjugation taking place between adjacent cells of the same filament (Fig. 32, *d*). Occasionally the two gametangia differ in size, and the conjugation tube of the one may be larger than that of the other, but no such differentiation of gametangia is found as in *Peronospora*. Nevertheless, green algae do exist, in which oogonia and antheridia are formed like those of *Peronospora*. We have abundant evidence, even in the few types here considered, that a differentiation of gametes into male and female sexes has taken place along many separate evolutionary lines in the plant kingdom. It appears to be a fundamental factor in the evolution both of plants and of animals.

CHAPTER X

FUCUS

Fucus is a brown seaweed which occurs in great abundance in the tidal zone all round the British coast, and related forms give rise to a similar "Fucus girdle" on the rocky shores of all temperate seas. It has a plant body which is fairly complex in structure and which grows to a maximum length of a few feet. As we have traced out the probable evolution of motile colonies and of simple filaments from unicellular green forms, so we may continue this evolutionary stringing together of plant-types, by deriving such plants as *Fucus* from filamentous forms of brown algae, these in turn having come from unicellular brown flagellate types.

Let us imagine the progression which possibly took place, and seek to deduce from the essential characters of the environment the main features which, almost inevitably, the evolving organisms must have developed. The evolution we are considering is that of *benthic* forms, i.e. forms attached to the sea bottom. We may imagine the brown flagellate settling down on the rock attached by its cilia, and we may imagine its descendants forming attached simple filaments. As the filaments grew in size, and as branching originated, the plant would begin inevitably to shew the effect of anchorage in the surf. The strain on the basal part of the filament would become very great, so that only forms possessing mechanical strength in this region would persist. Thus a holdfast part of the filament would evolve. In addition, the cells at the base of the plant would not be able to divide without weakening the axis dangerously (because

of the delicacy of newly formed walls), and so the tendency
would soon become evident for active cell division to go on
only at the apices of the filaments. Thus *apical growth*
would evolve. So long as the main axis was composed of
a *single* row of cells it could never become very strong, and
so the plant size would be limited. Such types we may
imagine superseded by others in which cell division in all
three planes, instead of one only, would give rise to solid
aggregations of cells, i.e. to "*parenchymatous*" structures or
to "*tissues*." These tissue plants would necessarily evolve
apical growth and holdfasts as before, but far greater possibilities of differentiation would be open to them. Thus they
might carry special assimilatory regions in the parts nearest
the light, special reproductive regions, and more complex
holdfasts. Further, in their solid cell masses, the outer cells
would be relatively well-nourished, and because exposed to
the light they would tend to become assimilatory cells, small,
and dividing rapidly, whilst the inner cells being relatively
starved would become colourless, and large, and often elongate. If the elongate inner cells should then become mainly
food-conducting channels, the plant would have achieved
"tissue differentiation," as well as the "differentiation of
members" shewn by the separation of holdfast, reproductive
regions, etc. *Fucus*, as the first tissue plant we have to deal
with, will shew all these characters we have deduced and
others of the same nature that time has not allowed us to
consider.

External structure of Fucus. The colour of *Fucus* is
brown; this is due to the presence in the plastids, in addition to the green chlorophyll, of an orange pigment,
fucoxanthin ($C_{40}H_{54}O_6$). Further, the proportions of the
four chlorophyll pigments differ from those of the ordinary
green plant, by there being relatively more of the carotin
and xanthophyll, and less of the chlorophyll α and β.

The structure of the body or thallus of the alga can be
seen from Fig. 33. It consists of a more or less cylindrical
stipe, or stalk, which is attached firmly to a rock or stone by

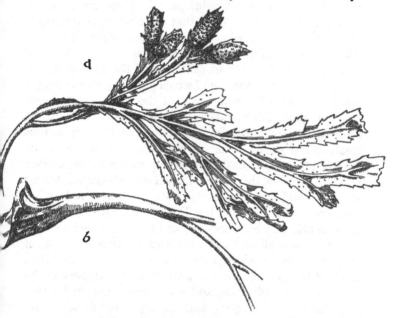

Fig. 33. *Fucus. a,* piece of the frond of *Fucus serratus* shewing the flat,
branched frond, and the rounded stipe derived from the thickened mid-rib of
the frond from which the wings have been worn away. The ends of four of the
branches are fertile, they are considerably swollen and contain numerous fertile
conceptacles which contain the gametes and shew on the surface as raised
papillae. Sterile conceptacles occur as dots scattered about the surface of other
parts of the frond. *b,* the base of a large *Fucus* plant shewing the large disc-
like holdfast detached from the rock.

a small flat disc-like holdfast, the cells of which are mucilaginous, and are closely cemented to the particles of the
rock. All the apical part of the plant is frondose, the *frond*
consisting of a flat, forked lamina, about $\frac{3}{4}''$ wide, with a

central thickened midrib. At the apex itself this midrib is inconspicuous, but as the frond gets older (as we pass to the base of the plant) the midrib gets thicker and at the same time the flat lateral wings of the frond become torn and beaten off by wave action, so that the old frond (or the base of the plant) has no wings, and is, in fact, merely the stipe. The flat frond ends in a narrow pit or groove, often containing short filaments of cells, and at the base of the pit are the cells by the division of which growth goes on. This is indeed the *apical growing point* of the alga. The terminal parts of some of the branches are swollen out into definite reproductive regions. On the surface of them numerous small raised papillae can be seen, each of which is the entrance to a small flask-shaped depression in the surface of the frond. In these flask-shaped hollows (conceptacles) the male and female gametes of the plant are produced in great numbers. Similar smaller sterile conceptacles also occur scattered about the rest of the thallus, and like the apical groove, all these conceptacles produce filaments of cells inside them which constantly exude mucilage. It is this which causes these seaweeds to be so slippery to the touch and to walk upon, and it has been suggested that it helps to prevent tangling and tearing of the fronds in the swirl of the rough surf. In some species of *Fucus* bladders are formed in the midrib and wing of the thallus by the splitting apart of the tissues, and these contain gas and serve to buoy up the alga when submerged. It is clear that we may readily make out in *Fucus* a differentiation of external members in accordance with specialisation of function in the directions of attachment, reproduction, assimilation, etc.

Microscopic structure. The structure of the cellular tissues of parenchymatous plants can best be studied by thin microscopic sections cut across them in various planes. From these a conception of the solid structure of the tissue can be gained. The most instructive sections are generally those at right angles to the length of the organ (*transverse* or cross-sections), but the two sets of *longitudinal* sections cut at right angles to each other should always be examined in addition.

Frond. A transverse section of the frond an inch or two from the apex shews two flat wing parts, and the swollen midrib. In each wing the cells form three clearly marked regions, the outer (palisade) layer of cells, the inner mass of cells (the medulla) embedded in a clear mucilaginous matrix, and the intermediate region (the cortex) between medulla and palisade tissues (Fig. 35).

The palisade layer is the chief photosynthetic tissue of the plant. The cells which constitute this layer are cylindrical cells elongated at right angles to the surface, and densely packed with the brown phaeoplasts which carry the chlorophyll pigments and fucoxanthin. The cells are all vacuolate and uninucleate, and have firm cell-walls containing a large amount of cellulose. They closely resemble in appearance the palisade cells of the higher land plants, save that they are not separated by air-spaces, and have no epidermal layer of cells over them. They are in direct contact with the sea water, from which they obtain supplies of carbon dioxide and mineral salts. The palisade layer is only one cell thick. Below it the cortex consists of four or five layers of cells of isodiametric proportions, packed closely together so that they appear polygonal. They contain far fewer phaeoplasts than the palisade cells and are usually regarded as constituting a storage tissue below the region of photosynthesis. Usually they shew a gradation in size from smaller cells outside to larger ones next the medulla, possibly an expression of the change in nutritional conditions at the outside and inside of the frond (see p. 162). These cortical cells play an important rôle, as we shall see later, in the formation of strengthening cells in the midrib, stipe, and holdfast. The cells of the medulla are grouped together in long chains, or filaments, which become separated by the mucilaginisation of all the longitudinal walls. The transverse walls remain thin and so these primary medullary cells appear as long threads wandering more or less parallel, but quite separate, in the mucilage of the medulla. Each cell contains some protoplasm, a large vacuole and occasional phaeoplasts, and has a thickened layer of cell-wall round the protoplast which has not been converted into mucilage. The direction of the primary medullary threads is different in the wings and in the midrib. In the latter they run strictly longitudinally (i.e. vertically), so that they are seen in transverse section as a number of small separate circular cells, but in the wings they run almost at right angles to this course (i.e. horizontally), and so in a transverse section of the frond are seen cut more or less along their length. The medullary threads of wing and midrib form one continuous system, which may be regarded, on account of their elongate nature and thin transverse walls, as the path by which organic food materials most readily move from one part of the plant to another (for example, from those parts of the fronds where photosynthesis is most active to all the growing points where cell formation is going on), and to the developing stipe and holdfast.

Primary growth. All the cells which we have so far spoken of, are formed by the apical growing point of the frond, and so are said to be of primary origin. The growth at the frond apex centres round the activity of a single large apical cell, in the surface layer of the cells of the terminal groove. This cell is a six-sided structure densely filled with protoplasm and its contents repeatedly undergo division. Mitosis is followed by the formation of a new cell-wall parallel to one of the four side walls or to the base of the

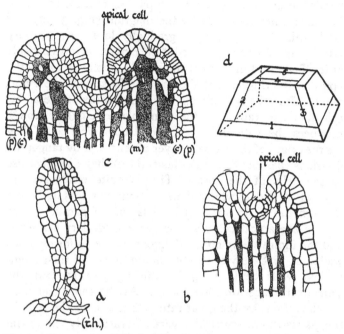

Fig. 34. *Fucus. a*, longitudinal section through a very young plant shewing the first signs of mucilage formation by the cell-walls. The apical groove and rhizoids (*rh.*) are just forming. *b*, longitudinal section through slightly older plant shewing the apical cell, further mucilaginisation of cell-walls and elongation of the central cells. *c*, longitudinal section through the apex of an adult frond shewing cell division proceeding at the apical cell and differentiation going on into palisade tissue (*p*), cortex (*c*), and medullary filaments (*m*). *d*, diagram of the apical cell shewing the successive planes of division numbered from 1 to 5. In all diagrams mucilage is shewn by shading. (After Oltmanns.)

apical cell, and at each division a new flat cell is thus formed, whilst the apical cell grows to its former size. In rapid and regular succession cells are cut off parallel to these five walls, in the manner indicated by the diagram (Fig. 34, *d*), and whilst all the cells cut off gradually lose their capacity for cell division, the apical cell itself remains always capable of it. The rapid division of the daughter cells cut off by the apical cell brings about the increase in size of the *Fucus* plant. The cells cut off from the lateral walls of the apical cell divide so rapidly in the same plane that their growth produces the walls of the pit round the apical cell. They are the origin of the palisade layer of cells, and by tangential divisions (parallel to the thallus surface) they also give rise to the cells of the cortex. This is quite easily seen in Fig. 34, *b* and *c*, which shew sections cut longitudinally through the frond apex. The cells cut off parallel to the *base* of the apical cell give rise to the medullary filaments, and the figure shews how, although they are uniformly thin-walled at first, very soon the middle layers of their longitudinal walls begin to grow mucilaginous, so that the separate threads are forced apart. At the same time they are drawn out by the great rate of increase of the cells of the palisade and cortical layers. Occasionally the apical cell divides longitudinally into two equal cells, each of which continues to behave as a separate apical cell. The cells produced by this activity form two fronds, which, as they grow, diverge continually from each other, in the manner seen in any adult frond.

Secondary growth. *Fucus* shews an extraordinary method of increasing the thickness and strength of the midrib and the stipe, a method which, because it involves cell growth from places other than the apical growing region, is called a kind of *secondary* growth. Although the midrib of the *Fucus* frond is exactly the same structure in origin

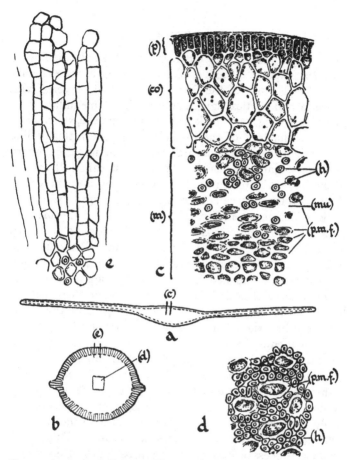

Fig. 35. Anatomy of *Fucus*. *a*, diagram of transverse section of the frond shewing the swollen mid-rib and the region (*c*) from which the high-power drawing *c* was made. The area within the dotted lines is the pith or medulla. *b*, diagram of transverse section of the old stipe shewing the remains of the two wings and shewing the great increase in thickness undergone by the mid-rib from stage *a*. The shaded part shews the zone of secondary thickening, and (*d*) and (*e*) shew the regions from which the high-power drawings *d* and *e* were made. *c*, transverse section of the young frond shewing outer palisade (photosynthetic) tissue (*p*), cortex (*co*), and medulla (*m*); the primary medullary filaments (*p.m.f.*) are widely separated by the mucilage (*mu*) produced from their cell-walls. Into this mucilage numerous thick-walled hyphae (*h*) have grown from the cortical and medullary cells. *d*, transverse section of part of the medulla of the stipe shewing the great contribution to its thickness and strength made by the increase of the hyphae between the primary medullary filaments. *e*, transverse section of the outer part of the stipe shewing the radial arrangement of the cells produced in secondary thickening.

as the stipe, yet it is very much easier to break in the fingers than the stipe, and this peculiar thickening is the explanation. In all parts of the frond the midrib is covered with palisade and cortical layers. Cortical and medullary cells give rise to long, very thick-walled, tubular cells, which grow into the mucilage of the medulla, and penetrating it like fungal hyphae grow downwards in it all amongst the primary medullary filaments. Such cells are best called "hyphae." They are readily recognisable in cross-sections of the young frond as a layer of cells with very thick, brightly refractive walls, widely scattered in the outer layers of the medulla just below the innermost cortical cells (Fig. 35, *a* and *c*). Throughout the length of the midrib, and all through the development of the alga, these hyphae are produced and grow down into the medulla. Successive transverse sections of the midrib, lower and lower down the plant, shew that the hyphae take up a greater and greater part of the space between the original medullary filaments, until in the old stipe the medulla seems a solid mass of bright, thick-walled hyphae, densely crowded together, and among which at intervals the browner, more open cells of the primary filaments can be made out (Fig. 35, *d*). These descending hyphae, reaching the base of the stipe, spread outwards and apply themselves closely to the rock, and so increase the area and efficiency of the button-like holdfast. This extraordinary process of secondary growth, seldom seen elsewhere in the plant kingdom, has the great advantage that, as the plant increases in size, by this means the supporting organs gain proportionately in strength.

The transverse section of the adult stipe shews the hyphal method of thickening to be supplemented to some extent by another method. The cells of the outermost cortical and photosynthetic layers begin to carry on active cell division, especially in a plane parallel to the surface of the stipe, and this produces a definite radial arrangement of the new cells which form a kind of cortical zone round the older tissues (see Fig. 35, *b* and *e*). Similar methods of increase in thickness of adult stem structures are very frequent among the more complex land plants.

Reproduction. When a *Fucus* plant is broken up by severe wave action, the pieces will float about alive for a

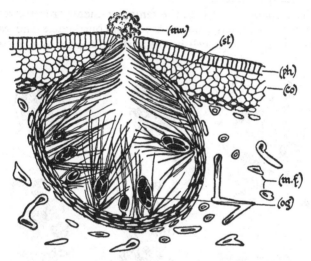

Fig. 36. *Fucus.* A conceptacle as seen in a transverse section through the fertile part of the thallus of *F. serratus.* The conceptacle is entirely female and contains several oogonia (*og*) and very numerous sterile filaments (*st*). From the narrow mouth of the conceptacle mucilage (*mu*) can be seen exuding. In the thallus note the photosynthetic layer (*ph*), cortex (*co*) and medullary filaments (*m.f.*) embedded in mucilage.

considerable time, but they never become attached again. Thus there can be said to be no effective vegetative reproduction, and there is no asexual reproduction. Sexual reproduction takes place by quite highly differentiated gametes produced in the fertile conceptacles at the ends of the fronds (Fig. 36). The differentiation of sex extends in some species to the plants, which then bear only male or female gametes; in some cases separate fronds of the thallus are male and female, and in other cases both male and female gametes are produced within the same conceptacle. The appearance of the thallus is not altered by the type of gamete it produces.

The oogonia, in which the female gametes are produced, are present in large numbers in the conceptacle and each arises from a single cell of the surface layer of the concep-

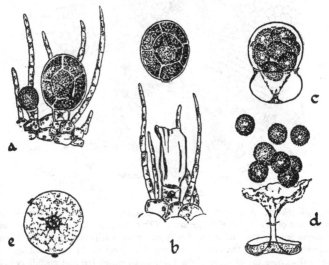

Fig. 37. *Fucus. a*, the origin of two oogonia on the inner wall of the concep-tacle, the one on the left is very young and the other already contains eight eggs. *b*, the outer wall of the oogonium has broken and set free the eggs within the middle and inner walls. *c*, the middle oogonium wall is turning back like a glove, and in *d* has turned almost inside out; the inner wall has broken down and has set free the eight eggs. *e*, shews the nucleus of a sperm passing through the egg-cytoplasm towards the egg-nucleus with which it will shortly fuse; two other sperms are seen outside the egg. (After Thuret and Farmer.)

tacle. This cell extends, and divides to form a short two-celled filament, of which the upper cell becomes large and rounded, forming the oogonium. In it, the nuclei divide three times, and round the daughter nuclei the cytoplasm aggregates to form eight large egg-cells closely packed together. The oogonium wall consists of three distinct layers, which are all ruptured, at different times, before the eggs are set free. This regularly occurs when the first waters of the rising tide cover the plants again after the period of exposure and drying at low tide. The middle layer of the oogonium wall first begins to become mucilaginous and to swell, so that the outer layer bursts and sets free the eggs in a bladder composed of the inner and middle layers. At the same time a great deal of mucilage is also produced by the numerous sterile filaments of cells present in the conceptacles, so that the bladders of eggs are exuded upon the surface of the frond in a mass of mucilage (which is olive-green in colour when produced solely from female conceptacles). The middle oogonium wall begins to turn back like a glove from the bladder, and when it has turned quite inside out the inner oogonium wall softens and the eight eggs are set free in the sea water. They thereupon separate and become quite spherical (Fig. 37).

The *antheridia*, which produce the male gametes, are produced in a closely similar manner to the oogonia, but they occur far more profusely on all parts of the large branched filaments of cells, which grow from the surface cells of the conceptacle. The antheridia are far smaller than the oogonia, and each is a club-shaped cell within which, following nuclear division, sixty-four male gametes (sperms) are produced. These closely resemble the male gametes of *Volvox*, for each is a naked, motile cell with an eyespot, and two cilia attached laterally (Fig. 38, *e*). The *Fucus* sperm is much rounder, however, and contains a single orange phaeoplast, whilst the *Volvox* sperm has a more or less

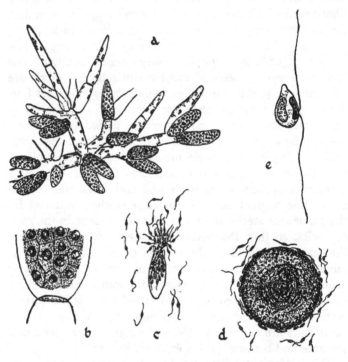

Fig. 38. *Fucus. a,* branched filament taken from a conceptacle and bearing ripe antheridia containing the male gametes (sperms). The antheridium has a three-layered wall and escapes from the conceptacle still enclosed in the two innermost layers; the outermost layer remains on the filament. *b,* the base of an antheridium which is almost ripe shewing the thick mucilaginous wall and the single sperm in formation inside each constituent cell. *c,* a single antheridium opening in the sea water outside the conceptacle and liberating the sperms by the rupture of the two inner antheridium wall layers. *d,* numerous sperms swarming round a fertile egg prior to its fertilisation by one of them. *e,* a single sperm shewing nucleus, phaeoplast, and laterally inserted cilia. (After Thuret and Guignard.)

degenerate chloroplast. Exactly as in the oogonium, the antheridium wall consists of three layers, and when the fronds are wetted by the rising tide the sperms are ejected in a bladder composed of the two inner layers. These bladders are pushed out upon the surface of the frond in a mass of mucilage which is coloured orange by the phaeo-plasts of the sperms, and with the rising of the tide both the inner and middle layers of the antheridial wall dissolve, setting free the gametes (Fig. 38).

Thus, as the tide covers the *Fucus* belt, large numbers of eggs and sperms occur together in the sea water. The eggs are non-motile but they secrete some substance which causes the sperms to swim towards them, and in microscopic preparations the sperms can often be seen to cluster so numerously round an egg that the lashing of their cilia sets it spinning round. Eventually one of the sperms penetrates the egg and the male and female nuclei fuse (Fig. 37, *e*). The considerable disproportion in size between the gametes is due to the abundance of food reserves in the egg and the lack in the sperm; both contribute equally to the nuclear matter bearing the hereditary characters of the plant. The zygote, when formed, germinates at once to form a new plant; a cell-wall is formed at right angles to the light falling on the cell, and a division into two cells takes place of which the one farthest from the light forms a simple branched filamentous holdfast, which immediately attaches the embryo plant to the substratum in which it has settled. The cell nearest the light divides repeatedly to form a little cylindrical pillar of cells, four or five thick, with a single apical cell. This body soon becomes flattened, and resolves itself into a small frond with a midrib and apical groove. After this, development proceeds along the lines already indicated, to the production of the adult *Fucus* plant.

Relation of Fucus to its life conditions. Unless the *Fucus* plant were adapted as a whole to life under the rather special conditions of the tidal belt of the shore, it would of course cease to exist there. We have already suggested some of the particular features which, in the structure or life history of the alga, seem to shew a close relationship to the environment, and we may now summarise and supplement them. In structure the thallus possesses great mechanical strength, and there is considerable division of labour both in the tissues and in external organs. Assimilatory, reproductive, attachment, supporting and growing regions are differentiated, and assimilatory, storage, mechanical, conducting, dividing and gamete-forming tissues. This complexity has given a plant superior in organisation to simple filamentous forms. It is doubtless successful in competition with these smaller forms by overshadowing them, and depriving them of light for photosynthesis. This tendency for larger and larger forms to evolve, by competition for light, has probably been responsible for the enormous seaweed forms of the South Atlantic (e.g. *Macrocystis*, up to 60 metres long), and probably also for the large size of forest trees on land (e.g. *Sequoia*, over 100 metres high). A particular factor of the environment with which *Fucus* must fit is the periodic immersion in water and exposure to the air. *Fucus* can often be found bone-dry on the rocks at low tide on a summer's day, but it has the capacity, shared also by a few other algae, of complete recovery when again placed in water. The reproduction of the alga must also be closely related to these conditions, for as we have seen, the release and fertilisation of the gametes depend on the tidal rhythm. The gametes are no more differentiated than those of *Volvox*, and probably the absence of all means of reproduction, save by them, is related to the efficiency of their dispersal in the tidal water. We may certainly suppose that the release of naked egg-cells into the water, and their germination, when fertilised, without any rest period, are connected with the general uniformity of conditions in the sea, and the unlikeliness of prolonged adverse circumstances. As in the green alga *Spirogyra*, so in *Fucus* we may suppose that the surface film of soft mucilage is of particular value in preventing the attachment and growth of small algae (epiphytes) such as we see making a heavy and continuous carpet on submerged structures such as pier supports and the stones of breakwaters.

It is interesting to observe that the mucilage which plays so large a part in the structure and biology of the brown sea-weeds has assumed a special economic importance during wartime. A considerable industry has grown up for collecting these sea-weeds and using them for the production of alginic acid, modifications of which yield such varied substances as textiles, custard-powders, and nutrient medium for bacterial cultivation.

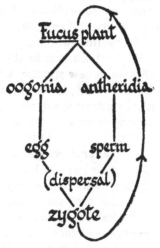

Diagram of life-cycle of *Fucus*.

CHAPTER XI

FUNARIA AND PELLIA

FUNARIA

In the types of the colonial and filamentous green algae, and in *Fucus*, we have considered chlorophyllous plants which are surrounded by water containing in solution the carbon dioxide and mineral salts essential for assimilation and growth. At some stage in the history of the evolution of land plants, such types as these must have given rise to simple organisms which colonised the dry land, and became the ancestors of all the present races of land plants. In the plants now fully established on the land the problems of structure must be very different from those of the algae; the medium no longer supports the weight of the organism, and no longer are mineral salts so readily available; desiccation is an ever-present possibility, and reproduction can no longer take place exclusively by free-swimming gametes. These, and innumerable other modifications in the environmental factors, must be reflected in new structures, tissues, and reproductive processes in the land plants. All such innovations must have originated by alterations in characters present in the body of the ancestral aquatic forms. Traces of this may still be seen in many land plants, and some scattered aquatic characters may be seen unaltered even in very highly evolved land plants (e.g. motile male gametes).

The steps of transition between the algae and the land flora are yet unknown, the fossil record having yielded no clear connecting links, though in one geological age (the Silurian), only algae have been found as fossil plants, and in the next (the Devonian), undoubted land plants have been identified. In the absence of definite palaeontological knowledge, we may now consider the structure and life-cycles of the simplest *existing* land plants, looking to find in them the same type of transitional characters that must have existed in the absent fossil links. One of the simplest of such existing plants is *Funaria*, the cord moss. It belongs to a group, the *Bryophyta*, including liverworts and mosses, in which the complexity of the plant body and reproductive organs is intermediate between that of the algae and of the higher land plants. The situations in which the *Bryophyta* live are as a rule very moist; thus liverworts and mosses occur very commonly in the undergrowth of woods, and by the sides of streams, where soil and air remain moist, and the chances of drying out are small. These plants have little in the way of protective coverings, but like *Fucus* they often retain the capacity for completely drying up and recovering uninjured when wetted again.

Many of the liverworts have a flat frondose plant body of the same degree of external differentiation as that of *Fucus*, and similarly called a thallus. In the mosses, however, external organs, superficially comparable with those of the seed plants, are developed. There is a main aerial axis (stem) bearing small ovate leaves in a spiral upon it, but no true roots are present of the complex multicellular structure of those found in seed plants. The roots of the seed plants are represented in the moss by long filaments of cells called rhizoids, which arise from the outer layers of the stem, and by penetrating the substratum serve at once as a means of anchorage and as a means of absorption of the dilute soil solutions. Evaporation of water from the leaves brings about a corresponding water intake by the rhizoids, and thus a continuous stream of water passes through the plant from the soil into the air. This "transpiration stream" is the physiological process most characteristic of all true land plants, in the lives of which it plays, as we shall see in the following chapters, a very important part.

Funaria is a moss which commonly occurs in close green tufts upon damp walls, paths, and especially places where wood fires have recently been made. The tufts are dense aggregates of more or less separate, sparsely branched *Funaria* plants. The upper parts of the stem and the upper leaves are bright green, but the lower parts, cut off from the light, are brown, having lost their chlorophyll. As each stem branches and extends, the older portions constantly die away so that axes which originated as branches now produce rhizoids and continue existence as separate units. In this very simple and effective way the *Funaria* reproduces vegetatively and the tufts constantly increase in size.

Structure. The leaves are very simple in structure, consisting of a flat, ovate plate with an acute apex and a broad base of attachment to the stem. The plate is one cell thick except for a thickened midrib in which the uniform brick-like cells of the rest of the leaf are replaced by a strand of narrow elongated cells, such as are found in the middle of the stem (Fig. 39). These probably afford the route of conduction of water and organic food in solution, through the leaf and between the leaf and the stem. The cells of the flat thin blade of the leaf are densely filled with ovoid chloroplasts which may, in growing leaves, be observed multiplying by division. No waterproof layer (cuticle) is present over the surface of the moss leaf, as in the higher plants, but water is taken up by the leaves and lost from them with great readiness.

The stem of *Funaria*, like the frond of *Fucus*, has a single apical cell from which the cells of the whole axis are formed. This is a three-sided pyramidal cell with the base outwards, and divisions occur in sequence on the three pyramidal faces. The cells so cut off divide by a wall parallel with the outside of the stem, and of the two cells so formed, the inner forms especially the stem tissues, and the outer gives rise to the

leaves and the buds from which new lateral stems arise
(Fig. 39, *f* and *g*). In the mature stem we can distinguish

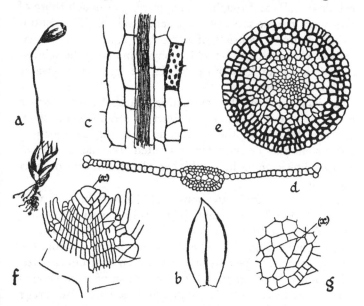

Fig. 39. *Funaria. a,* the whole plant shewing the short axis bearing leaves
and rhizoids. Growing from the stem apex is a mature sporogonium consisting
of a pear-shaped spore capsule on a slender stalk. *b,* a single detached leaf.
c, surface view of part of a leaf shewing the midrib of elongated cells running
through the green assimilatory cells forming the rest of the leaf (one shewn
containing chloroplasts). *d,* transverse section of a leaf. *e,* transverse section of
the stem shewing the outer cortex of thick-walled cells and the central (conducting) strand of thin-walled cells. *f,* diagram of longitudinal section through
the apex of the stem of the water-moss (*Fontinalis*) shewing the method of
growth from an apical cell (*x*). *g,* diagram shewing transverse section through
the apex of the stem of *Funaria,* and passing through the three-sided apical
cell (*x*). (*f,* after Leitgeb, and *g,* after Campbell.)

three distinct regions, which may be called outer cortex,
inner cortex and central conducting strand. The cortex cells

when young contain chloroplasts, and in both zones of the
cortex the cells are parenchymatic (i.e. living cells with thin
cellulose walls and roughly isodiametric), although those of
the outer cortex are usually thicker walled. The central
strand, however, consists of narrow, elongated, thin-walled
cells. They contain no chlorophyll and like the similar
strands in the midribs of the leaves, are regarded as a water
conducting tissue.

In *Funaria* itself, the conducting strands from the leaves
do not fuse with the main conducting strand of the stem,
but end blindly in the cortex, so that the efficiency of
the system may be regarded as limited. In other species of
mosses, however, as in the higher plants, the continuity is
complete.

The moss rhizoids are branched filaments of cells with
oblique end walls, and although generally brown where below
ground, they develop chloroplasts in places where they come
into the light (Fig. 42). The diameter diminishes from the
base towards the apex so that the system is of limited size.
Although far simpler than the root system of higher plants
it undoubtedly carries on as they do the functions of anchorage
and absorption.[1]

Thus in somatic structure the moss shews adjustment to
terrestrial conditions. The possibility of desiccation, the small
contact with the food medium, the absence of support by
the medium, are all problems which have been successfully
met, by methods involving some degree of differentiation
beyond that of an alga like *Volvox* or *Fucus*.

Reproduction. The reproductive processes of the simple
land plant must similarly have to meet problems not present

---

[1] This may be verified by leaving the plants of *Funaria* for an hour or two
standing with the rhizoids in an aqueous stain such as eosin, or Delafield's
haematoxylin, and the paths of water conduction will be found, on cutting
sections, to have taken up the stain to a greater extent than the other tissues.

to the alga. Thus the gametes and zygotes will both be less readily dispersed than in the liquid medium, and gametes, zygote, and embryo, may meet much more adverse, and more extreme conditions, especially desiccation. In the increased complexities of the reproductive mechanism we may expect to find the effects of this.

The sexually differentiated gametes of *Funaria* are borne within antheridia (male) and archegonia (female). In each case, a wall of cells surrounds the developing gametes, and only at maturity do they come into contact with the surrounding atmosphere. The antheridia arise in groups at the apices of the stems, in a cup-like rosette formed by the foliage leaves; they are interspersed with large sterile filaments the apical cells of which are much swollen (Fig. 40, *a, b, c*). The antheridia are club-shaped and inside the wall there are very numerous small cells each of which gives rise to a single motile sperm. The walls of these cells become mucilaginous, and the sperms are extruded from the apex of the antheridium and may be carried in rain water to the female organs, which are borne on lower branches of the same plants. These female organs, the archegonia, are flask-shaped bodies consisting of a neck and a lower part (the venter). The neck consists of seven vertical rows of cells, six being peripheral and the other (canal cells) central. Just before maturity the venter wall becomes two cells thick, and it may be seen to contain a single large egg-cell and a ventral canal cell, which lies between the egg and the neck canal. At maturity the apex of the neck opens, the canal cells and the ventral canal cell break down, filling the neck with soft mucilage. The sperms, having fortuitously reached the vicinity of the neck, swim down it and one of them fuses with the egg. It has been established that there is a chemotactic response by the sperms to the secretion of cane sugar by the egg (cf. *Fucus* and *Volvox*).

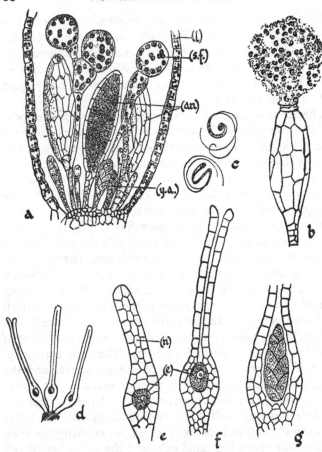

Fig. 40. *Funaria*, development of the gametes. *a*, longitudinal section through the apex of a shoot bearing mature antheridia (*an*) and young antheridia (*y.a.*), among large sterile filaments (*s.f.*) which are club-shaped; (*l*) leaf. *b*, a ripe antheridium from which ripe sperms (male gametes) are being protruded in a mass of mucilage. *c*, the mature coiled sperms with cilia. *d*, group of archegonia from the apex of a female branch. *e*, young archegonium shewing the egg-cell (*e*) and intact cells in the neck (*n*). *f*, ripe archegonium with neck open to sperms. *g*, early divisions of the fertilised egg-cell within the archegonium. (*a* and *b* after Sachs, *c* and *e* after Campbell, *g* after Baines and Land.)

Development of Zygote. The zygote does not, as in *Fucus*, germinate directly into a new plant, but proceeding to divide inside the venter of the archegonium (Fig. 40, *g*), gives rise to a new structure, a "sporogonium." This is a complex body, inside which is produced a very large number of small wind-dispersed spores, which serve to spread the moss plant, since they give rise, on germination, to a green alga-like body (protonema), which bears new plants of *Funaria*. The sporogonium consists of foot, stalk and capsule. The foot is embedded in the apex of the leafy moss stem and the capsule is a pear-shaped body which droops at the end of the thin wiry stalk (Fig. 39, *a*). The whole is perhaps 3 cms. high, and shews a complexity of structure with which space does not permit us to deal at all fully. The base of the capsule is green in colour, and sections through it shew that it has a definite epidermal layer of cells, covered by a thin cuticle. This is pierced at intervals by numerous stomata, and below it occur several layers of green assimilating cells densely packed with chloroplasts, and separated by wide air-spaces (Fig. 41). This structure is that of a thoroughly efficient organ for carrying on photosynthesis whilst raised into the air. That is to say, it obtains carbon dioxide from the small concentration of that gas present in the air, and at the same time, the absorbing cells are protected against desiccation by the dry air. In this it can be said to foreshadow almost all the chief characters of the photosynthetic region of the green leaf of the higher land plant. The sporogonium stalk contains a central strand of elongated cells, doubtless playing much the same rôle as the similar cells of the main moss stem. Only the upper part of the capsule is fertile, and the spores are produced upon a central barrel-shaped "columella" of sterile tissue, which is separated by a wide air-space from the capsule walls. When the capsule is ripe, most of the tissues within it collapse, and it remains filled with a

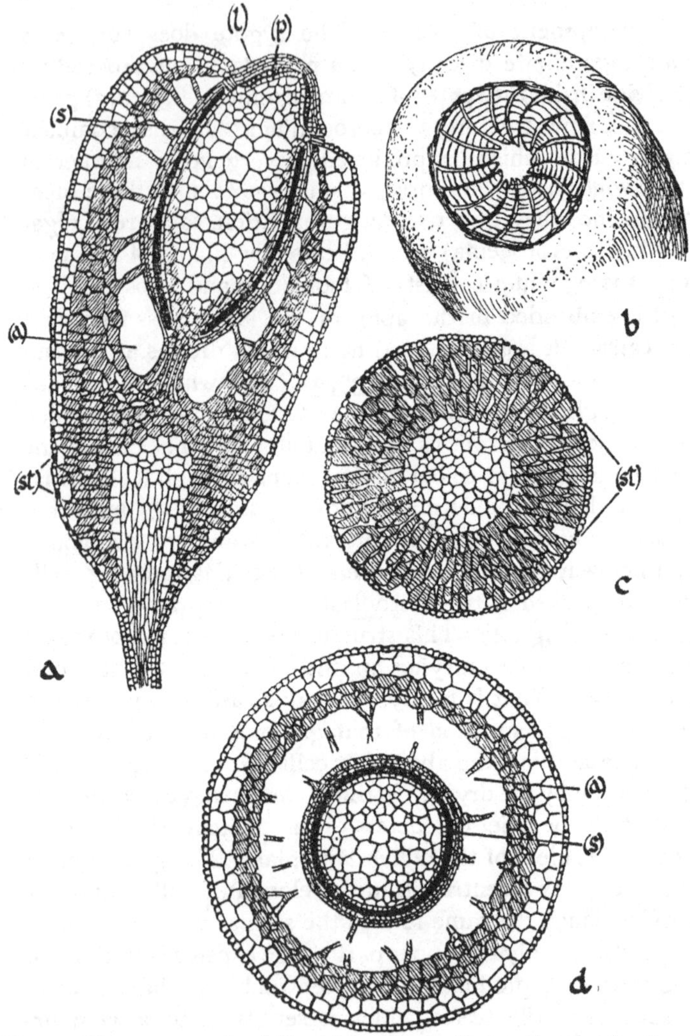

Fig. 41. The spore-capsule of *Funaria*. *a*, longitudinal section shewing the lid (*l*), the spore-forming tissue (black) (*s*), the green assimilatory tissue (shaded) which occupies a great part of the lower part of the capsule where it is in contact

mass of dark green spores. The definite lid which closes the capsule now falls off, revealing below it a double row of long tooth-like bristles which project over the open mouth of the capsule (Fig. 41, *b*). These teeth are very hygroscopic, and close the opening when the air is moist, but in dry conditions curl back and allow the spores to be scattered. The spores contain chlorophyll and reserves of food material. They germinate to produce a branched filamentous green structure resembling a simple alga. This is the "protonema." It spreads about the surface of the soil often forming a tangled felt, and young moss plants arise from it vegetatively as small buds, in which an apical cell is formed at once (Fig. 42). This filamentous stage in the moss life-cycle emphasises the connection with alga-like ancestors, but the possession of wind-dispersed spores is a stage towards the complexity of the life-cycle of the higher plant, a complexity almost entirely conditioned by the limited water supply available. It would require far greater space than that available here to explain this statement adequately, and it must now suffice to say that the production of pollen grains which are carried by wind or insects, the process of pollination of the receptive part of the female organs of the flower, the production of seeds within a definite fruit, and the formation of the embryo-plant within the seed, and all the innumerable modifications of these principles, have followed from the terrestrial exist-ence of the flowering plant. Later chapters of the book must shew that in vegetative structure as well as in repro-ductive structure, these flowering plants are adapted most

---

with the outer air via air-spaces and stomata (*st*). Threads of chlorophyllous tissue also cross the wide air-space (*a*) round the central column of tissue. At the base the capsule narrows to the stalk in which the cells are elongate and colourless. Below the lid are the peristome teeth (*p*). *b*, capsule seen from above after the lid has fallen off, shewing the ring of peristome teeth whose movements control the escape of spores from the capsule. *c*, transverse section through the base of the capsule; *d*, transverse section through the upper part of the capsule. (*a*, after Haberlandt.)

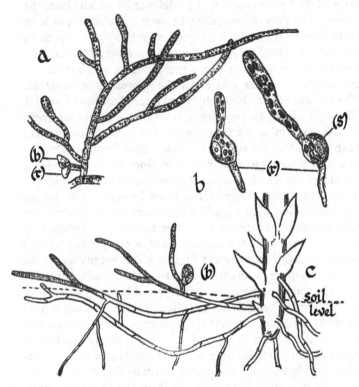

Fig. 42. *Funaria. a*, the primary protonema consisting of a branched filament of green cells; this gives rise to the ordinary moss plant by development of a bud (*b*). Many such may occur on one protonema. (*r*), a young rhizoid. *b*, spores from the moss capsule germinating to produce the protonema. Each shews one filament with chloroplasts and one without, the latter is the first rhizoid (*r*); *s*, the old spore wall. *c*, diagram of the base of a moss plant shewing the colourless or brown rhizoids produced below soil-level; where they come into the light they become green (indicated by shading) and so form a *secondary* protonema, which bears buds like the *primary* protonema which is developed from the moss spores. (After Thoday and Luerssen).

closely to life on land, a fact to which undoubtedly they owe the supremacy which they there enjoy over other forms of plant life.

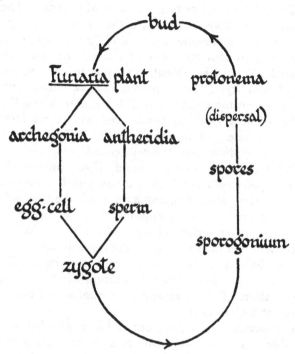

Diagram of life-cycle of *Funaria*.

PELLIA

A plant very strongly resembling *Funaria*, both in structure and life-history, is the liverwort *Pellia*, which grows closely above water-level on shaded stream-banks, and in similar moist places. The green part of the plant shews very little structural differentiation, for it consists of a thin green thallus, some half-inch wide, lying upon or close to the ground. A transverse section shews this to consist of very few rows of cells in the central thickest part, and it is but one or two cells thick at the edge. These cells are remarkably similar, save that those on the upper surface contain more chloroplasts, and that in surface view it can be seen that the cells of the mid-rib are rather elongated. Cuticle is absent, and the fronds are very susceptible to drying. Each frond of the green thallus ends in a small growing-point set in a depression, where there may be found a small pyramidal cell with three lateral faces and a base, just like the apical cell of the *Funaria* stem. It is by division of the cells cut off from this apical cell that the new thallus is formed. From time to time two new apical cells are formed by an equal longitudinal division of the parent cell, and as a result the thallus has a characteristic bifurcating appearance much like that of *Fucus*. The under surface of the mid-rib of the thallus is covered with fine tubular threads, the rhizoids, and each of these can be seen to be an outgrowth of a single cell of the thallus surface. The rhizoids, apart from being unicellular, resemble those of *Funaria* and have similar functions of attachment and absorption.

Reproduction. A single plant of *Pellia*, growing under favourable conditions, soon multiplies by outward growth and branching of the fronds, coupled with decay of the older parts; this is purely vegetative reproduction. Equally however with *Funaria*, *Pellia* shews a life-cycle in which gamete formation with conjugation, and spore formation regularly succeed one another. The sexual organs are once more archegonia and antheridia and the spores are again produced in a sporogonium.

In spring the green thallus will often shew to the naked eye a number of small dark dots scattered near the mid-rib.

Each of these is an antheridium, extremely like those of *Funaria* in structure, although spherical in shape, and so deeply buried in the thallus that they communicate with the surface by only a small hole (Fig. 43). Each antheridium has a short stalk and a single layer of wall cells. Within, the protoplasm becomes divided up into a large number of spermatozoid mother cells, each of which produces one spermatozoid. In antheridia which are nearly ripe the coiled form of the sperms can be made out. The antheridium bursts open at maturity by the absorption of water and the extrusion of a mucilaginous mass of cells, from which the sperms eventually escape. Each spermatozoid is a spirally curved colourless mass of nuclear material, bearing two cilia, and capable of active movement.

The archegonia of *Pellia* are borne in a small cup which originates near the frond apex, and soon stands up a millimetre or two above the mid-rib. Within it are the flask-like archegonia, extremely similar to those of *Funaria*, with the same long neck, venter, egg cell, neck cells and ventral canal cell. The wall of the venter is, however, made of only one layer of cells. The opening of the neck cells and collapse of the neck-canal cells is followed by the entry of the spermatozoid, splashed by rain and swimming in surface films of water until it responds chemotactically to the secretion of the egg cell. Fertilisation of the egg cell gives rise to a zygote which at once begins to divide, and to form the sporogonium. It soon becomes evident that the sporogonium consists of a conical foot embedded in the thallus, a massive short stalk, and a globular head. The head and stalk for a long time remain enclosed in the archegonium wall which stretches and forms a protective covering, at the apex of which the archegonium neck remains easily recognisable. The head, that is the capsule, has a wall of two or three layers of cells and within it the cells are of two kinds, spore mother cells and elaters. Each rounded spore mother cell divides to form four ovoid spores. The elaters however are elongated tubular cells which develop marked spiral thickenings on their walls, and it will be seen that they are most abundant in the central basal part of the capsule (Fig. 43). As the spores become mature the stalk suddenly elongates and bursts through the calyptra. Within two or three days it extends from about one millimetre to forty or fifty times that length; this is merely due to cell-extension and vacuolation without any cell-division. This is a clear indication of the independence of the two processes of cell-division and cell-extension which so commonly are closely associated (cf. pp. 73–76). The sporogonium bursts through the calyptra, and in the drier air, into which the capsule is now raised, it splits open into four petal-like valves, and the spores are carried off by the wind. Their escape is assisted by the elaters which move hygroscopically as they dry, and which, as they coil and twist, jerk the spores from the capsule. The ripe spores shew a few cross-walls, i.e. they are multicellular, and they also contain chloroplasts. The spores have no powers of endurance and need to germinate at once. One of the cells of the spore forms the apical cell of a new *Pellia* thallus, and from others grow the first rhizoids.

Alternation of generations. From this description it will be clearly seen that *Pellia* resembles *Funaria* very closely in the general outline of the life-cycle. In both the conspicuous and independent phase is that which produces the gametes. This is referred to as the *gametophyte generation*: it originates with germination of the spore, proceeds through formation of the moss protonema and moss plant, or of the *Pellia* thallus, to the final production of sperms and eggs, respectively in antheridia and archegonia. Fertilisation initiates a second phase, which is more or less parasitic

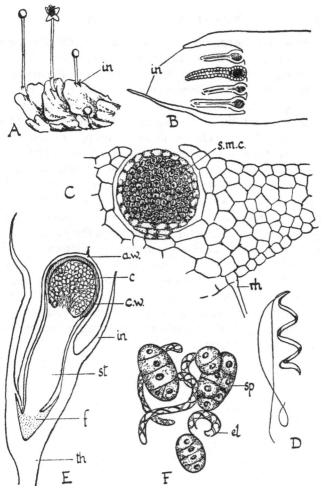

Fig. 43. *Pellia.* *A*, the prostrate thallus with sporogonia in different stages of maturity, each growing from a small cup or involucre (*in*). *B*, a group of archegonia within the involucre. *C*, section of the thallus with one rhizoid (*rh*), and a spherical antheridium embedded in the upper surface; the antheridium is packed with sperm mother cells (*s.m.c.*). *D*, a single spermatozoid. *E*, diagram of a young sporogonium within the involucre (*in*), and the old archegonium wall (*a.w.*); the sporogonium consists of foot (*f*) [embedded in the thallus (*th*)], the stalk (*st*), and capsule (*c*); *c.w.*, capsule wall. *F*, spores (*sp*), and elaters (*el*) from the capsule. (After Cooke, Scott and Guignard.)

upon the first, the sporogonium. This is called the *Sporophyte generation*, since it ends with the formation of spores.

The regular sequence of these two phases of the life-cycle is what is known as *alternation of generations*, and this can be recognised, not only in the Bryophyta, but in the life-histories of all the more highly organised plants. It is found with less regularity in lower groups of the plant kingdom. Both in the moss and liverwort the organisation of the gametophyte is simple, and though related to land conditions, shews strong affinities for moist places. The gamete liberation and conjugation also demand free water. The sporophytes of *Pellia* and *Funaria* are alike in their dependence on the gametophyte, although the *Pellia* capsule has no chlorophyllous tissue nor stomata. Both are clearly related to the dispersal of spores in dry air, the one with a lid and hygroscopic teeth, the other with valves and elaters. The moss protonema has no counterpart in the life-cycle of *Pellia* though many other liverworts have such a stage; it can be regarded as an early alga-like step in the development of the gametophyte.

CHAPTER XII

THE FERN PLANT

The phylum of plants to which the ferns belong is known as the *Pteridophyta*, and includes such types as the horse-tails, club-mosses, and many long-extinct species known only as fossils. This phylum shews, in comparison with the Bryophyta, a great increase in complexity of the plant body, together with increasing adaptation to life on land. As in the Bryophyta the life-cycle shews a strict alternation of generations, but in the Pteridophyta it is the sporophytic stage which is conspicuous and independent, whilst the gametophyte is small and often dependent upon the sporophyte for protection and nutrition. A short consideration of the organisation and life-history of the common British male-fern, *Nephrodium filix-mas* (also known as *Aspidium filix-mas* or *Dryopteris filix-mas*), will serve to illustrate most of these principles. A specimen of this plant dug up from the hedgerow at once reveals a plant body far more massive than that of any moss or liverwort. Each leaf is a long branching structure springing from just below soil-level two or three feet into the air, and divided into a great number of small pinnae; this is a striking contrast to the small size of moss leaves and the absence of leaves from such liverworts as *Pellia*. The stem in the male-fern is a creeping stem growing just below the ground surface, a so-called rhizome, which bears the leaves densely over its apical portion. The young leaves are tightly coiled in a spiral fashion, and only develop slowly, taking about three years to reach maturity. The rhizome apex shews the youngest leaves, behind come the stout petioles of the expanded leaves, and the rest of the rhizome is densely covered with the rotted bases of petioles of leaves now dead. The surfaces of the young petioles and rhizome are covered with thin brown protective scales. The rhizome apex can be shewn microscopically to grow forwards by the activity of a pyramidal apical cell similar to that of *Fucus* and of the Bryophytes. The rhizome branches from time to time, and as the old portions die away, vegetative multiplication results. In the bracken-fern the rhizome is slender and creeping; hundreds or thousands of feet of branching rhizome go to form one plant, and its great power of vegetative extension is no doubt partly responsible for its rapid and menacing advance as a weed into heaths, moors and pastures. From the rhizome grow out abundant branching *roots*: these are vastly more massive and complex in structure than the rhizoids of the Bryophyta. They have an apical growing-region protected by a *root-cap*, and a complex arrangement of internal tissues, much like the roots of flowering plants.

If the external organisation represents an increased complexity beyond that of *Pellia* and *Funaria*, still more does the internal differentiation of cell-structure. We cannot now discuss at all fully the microscopic anatomy of the fern, but a few special points may be shortly mentioned. The leafy anatomy much resembles that of the flowering plant already referred to on p. 88. There is an upper and a lower epidermal layer often containing chloroplasts, and the lower epidermis is pierced by stomata. The mesophyll is similar to that of the flowering plant but the palisade-tissue is less well-developed and it may all be of the spongy type with large irregular air-spaces (see Fig. 46). Through the mesophyll runs a coarse network of veins which are cut at different angles by the section, and which consist

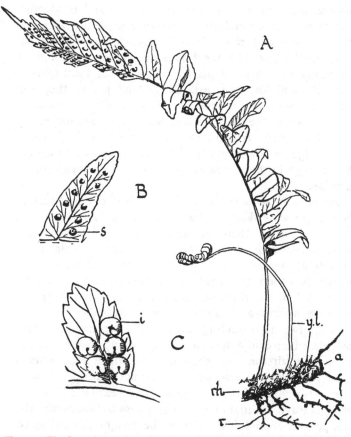

Fig. 44. *The fern*. *A*, portion of a plant of *Polypodium vulgare*, a common British fern, with creeping rhizome (*rh*) covered with scales, bearing roots (*r*) and young coiled leaves (*y.l.*) near the apex (*a*): the single mature leaf has fertile pinnules at the top, bearing sori. *B*, lower surface of a fertile pinnule of *Polypodium* shewing the veins and groups of sporangia, i.e., sori (*s*). *C*, lower surface of pinnule of the male-fern, shewing five sori, each covered with a kidney-shaped indusium.

of the complex tissues with mechanical and conducting functions that extend thence down the petiole, and into the rhizome and roots. The leaf-cuticle is thin and not very efficient, so that the leaves are liable to wilt and die in sunny and exposed places. It is, in fact, most marked that although the ferns are clearly land plants, they still have a preference for moist conditions. This is especially well shewn in the British Isles, where ferns are extremely abundant in the Atlantic western counties both in woods and hedgerows, whereas in the drier eastern part they are few enough in the woods, and almost absent from situations outside.

Associated with the large leaf-surface and root-surface is a large water-loss to the air and a rapid transpiration stream. It is not therefore surprising to find an elaborate water-conducting system, and a similar organic food-conducting system linking leaves, rhizome and roots. This, which is called the *vascular system*, is most strikingly developed in the rhizome, and can be displayed there by soaking in alkali a portion of rhizome divested of leaf-bases, and afterwards dissecting away the softer tissues. There will then be found a system of tough strands like a cylinder of coarse wire-netting, each wire of which represents a vascular strand. Each mesh of the net corresponds with one leaf-base, and from the strands round each mesh, branch strands leave the main cylinder and pass outwards into the leaf petiole. Something of this will be recognisable in Fig. 45. Each strand of the vascular network contains specialised tissues of different kinds, but of these we need stop to consider only the largest and most evident. This is the water-conducting tissue made up of xylem-tracheids, which can be examined in a portion of vascular strand macerated in strong alkali and then teased apart on a microscope slide. The largest structures then visible are tubular cells

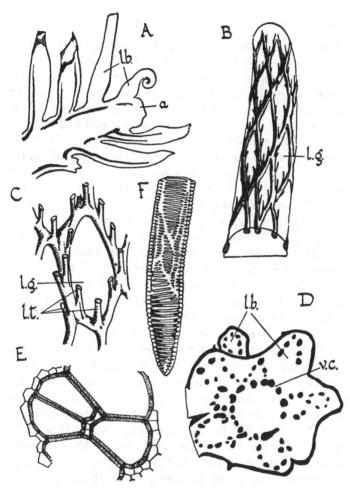

Fig. 45. *Nephrodium filix-mas*: structure of the sporophyte stem. *A*, long section through the rhizome, shewing the apex (*a*), and leaf-bases (*l.b.*) with a prominent system of vascular strands. *B*, the vascular cylinder dissected from a rhizome; *l.g.*, leaf-gap. *C*, portion of the vascular cylinder enlarged, shewed a leaf-gap and, given off from its margins, the leaf-traces (*l.t.*). *D*, transverse section of the rhizome, shewing the vascular cylinder (*v.c.*) and numerous leaf-bases (*l.b.*). *E*, a group of tracheids seen in transverse section. *F*, part of a single tracheid, shewing the ladder-like arrangement of pits and the wedge-shaped end. (After Sachs and De Bary.)

with several flattened faces and flat wedge-shaped ends. These flat faces are the surfaces at which the neighbouring tracheids meet in a closely packed tissue. The walls are thickened with strengthening material (lignin), but on each face between one tracheid and the next is a row of oval areas, or pits, where thickening is absent and only the primary cell-wall separates the two cells (Fig. 46). Through these pits a free movement of water takes place, whilst the cells are still strong enough to resist collapse when the water in them is under tension. A few tracheids do not have the oval pits in this characteristic "ladder" arrangement, but have a spiral lignin thickening running the length of the cell. Although one or two mosses shew a slight specialisation of water-conducting cells, nowhere in the Bryophyta is there anything approaching this complexity. A similar specialisation applies to the other parts of the fern anatomy in stem, leaf and root alike. There are present also representatives of all the chief tissues of the flowering plant, and only slightly less highly organised.

Spore production. As the fern-leaves become adult they may be seen to be covered on the lower surface with brown dots (sori) which are groups of sporangia. In the male-fern each sorus is protected by a kidney-shaped protective scale (indusium). The sorus is at the end of a small vein and consists of a small cushion of tissue with abundant tracheids, on which are borne numerous small capsules, the sporangia. Each sporangium is a multicellular structure, in shape like a biconvex lens set on a slender stalk. It has a wall of cells one layer thick, and running three-quarters round the edge of the lens shape a row of these cells is strikingly different from the rest. The cells of this *annulus* are regular in shape and very much thickened on all but the outer wall: the other wall cells are thin-walled, including the elongated cells forming the remaining quarter of the edge

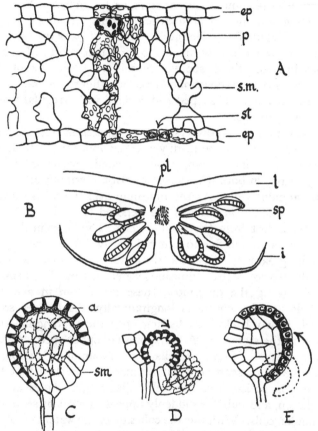

Fig. 46. *The fern.* A, transverse section through the leaf, shewing epidermis (*ep*), penetrated by stomata (*st*): the palisade (*p*), spongy mesophyll (*s.m.*) are not strikingly different. B, section of the leaf (*l*), bearing a sorus which consists of sporangia (*sp*) on a placenta (*p*), and covered by an indusium (*i*). C, a ripe sporangium containing spores (dotted), and shewing the conspicuous annulus (*a*), and thin-walled stomium (*sm*) where the opening begins. D, the sporangium pulled open by contraction of the annulus cells. E, the sporangium jerked back by formation of gas bubbles in the annulus. (A, after Bower, C after Campbell.)

between the annulus and the stalk, i.e. the *stomium* (Fig. 46).
As the sporangium approaches maturity its dense protoplasmic contents divide into a number of *spore mother cells*,
each of which finally divides to form four spores. In their
development these spores are said to make use of the
products of breakdown of other cells within the sporangium;
they become bean-shaped and covered with irregular wing-
like projections.

The process of dispersal of the spores is one of great
interest which can be readily watched under the microscope
by placing ripe sporangia on a heated slide or in strong
glycerine; in both it is caused by withdrawal of water from
the annulus cells. As the indusium over the sorus withers
the sporangia on the leaf are exposed to desiccation and
water is lost from the annulus cells. The pores in the wet
cell-walls are too fine to admit the air-water surface except
under great tensions, and so the annulus cells are contracted
under tension of the diminishing water content. The weakest
walls being the outermost, these are drawn in and the
whole annulus contracts longitudinally. As it does so it
curls slowly backwards, tearing open the sporangium wall
first at the stomium. A few spores fall out at once. Water-
loss continues from the cells until the annulus is strongly
reflexed, but soon so great a tension is reached that the
water is pulled from the cell-walls, or gas comes out of
solution, and bubbles suddenly appear in most or all of the
annulus cells. With this the cohesion of the water is broken,
and the whole annulus flies violently back to its original
position. The slow back swing tearing the annulus, and
the violent drive propel the spores considerable distances,
and upon the heated microscope slide the jerk will often
throw the sporangium from the microscope stage. The
purely mechanical character of the movement is shewn by
the fact that when the gas bubbles have disappeared in

solution from old opened sporangia kept in spirit, the movements can be readily repeated.

It is hardly necessary to emphasize the completeness of the adaptation to land conditions thus shewn by this phase of the fern's life-cycle. Many spores are produced in each sporangium, and the production of spores by even a single leaf reaches prodigious numbers. The spores are wind-borne and their prevalence is such that a handful of moist soil taken almost anywhere in Britain and sealed in a closed glass jar kept for some months or years in the light will not fail to produce young fern plants.

The Prothallus. The spores are capable of retaining life under dry conditions for some months, but when moistened they germinate to produce the small green *prothallus*, the gametophyte generation of the fern. This is a delicate green structure seldom longer than half an inch, and only a few cells thick in the stoutest central portion. It is heart-shaped, with a pyramidal apical cell in the terminal depression. On the lower surface abundant rhizoids attach the prothallus to the soil. The cells of this small thallus are almost undifferentiated, all being more or less isodiametric and containing chloroplasts. Although also auto-trophic in nutrition the gametophyte is much more closely dependent than the sporophyte on moist surroundings; cuticle is absent, and an hour or two in dry air shrivels it completely. In natural conditions prothalli only occur in permanently moist places, thus, for example, on dry heaths the bracken fern produces prothalli only in such places as the damp entrances to rabbit-burrows. On the lower surface of the prothallus the sexual organs are borne, the archegonia mostly on the thick portion next the apical depression, and the antheridia extending more towards the margins. The antheridia are spherical structures projecting from the surface; each has three wall-cells, two of which are ring-shaped, like a motor-tyre whilst the third is a disc-like lid cell (Fig. 47). The centre is occupied by sperm mother cells, each of which produces a single coiled spermatozoid bearing numerous cilia. The sperms are set free by the opening of the lid cell in wet conditions.

The archegonia resemble those of Bryophytes in their essential structure, but the venter is embedded in the tissue of the prothallus and the shorter curved neck contains only four rows of neck cells surrounding the canal cells. The mucilage formed by disorganisation of the canal cells contains malic acid, and this substance causes chemotactic response by the sperms, which swim towards the egg and fertilize it. For this, of course, some small amount of liquid water is essential.

The fertilisation of the egg *in situ* is followed by its germination, also *in situ*. It divides into four quadrants, which form respectively the stem, leaf, root and foot of the young sporophyte. The foot is a tissue embedded in the prothallus, and through it the embryo sporophyte takes food from the gametophyte. With the development of green leaves of increasing size and complexity, and of larger stem and more numerous roots, the sporophyte rapidly becomes independent and reaches the size and structure we have already described.

It is of great interest to note how the two generations of the fern respectively are adjusted to the moisture conditions of their habitat. The gametophyte can endure almost no drying, both antheridia and archegonia open by absorbing water, and the gametes need liquid water for fertilisation. The sporophyte, by contrast, has a complex external and internal organisation, in manifest relation to life on land, exposure to air, development of a transpiration mechanism, and utilisation of gaseous carbon-dioxide; the sporangia have a special dehiscence mechanism dependent on drying,

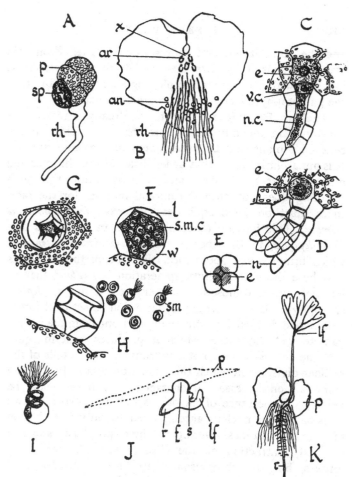

Fig. 47. *The fern.* A, young prothallus (*p*), first rhizoid (*rh*) and old spore (*sp*). B, prothallus, shewing apex (*x*), region bearing archegonia (*ar*) and antheridia (*an*). C, young archegonium with egg cell (*e*), ventral canal cell (*v.c*) and disorganised neck canal cells (*n.c.*). D, mature archegonium. E, archegonium seen from above. F, young antheridium with wall (*w*) and lid cells (*l*) and full of sperm mother cells (*s.m.c.*). G empty antheridium from above. H, the same in side view with sperms (*sm*). I, spermatozoid. J, young sporophyte with first leaf (*lf*), stem (*s*), root (*r*) and foot (*f*). K, young sporophyte with first leaf and root developed but still attached to the prothallus. (After Kny and Strasburger.)

and the drought resistant spores are air-borne. From this stage onwards it appears that evolutionary development has lain with the dry-land exploiting sporophyte, and that the gametophyte has become progressively smaller.

Nuclear Behaviour. We have already stressed the constancy in the number of chromosomes within all the cells of a plant and between plants of the same species. It is apparent however that at fertilisation the chromosome number must be doubled, and since we know that such doubling does not raise the normal number for the race, there must be some equivalent *reduction in chromosome number*. This halving takes place, both in *Bryophyta* and *Pteridophyta*, in the special *reduction-division* or *meiosis* which brings about spore formation. From this it follows that here the gametophytic generation is *haploid*, that is, has half the chromosome number present in the *diploid* sporophyte. Each sporophyte has a double set of chromosomes, one derived from the male, and one from the female gamete, but the spores which it produces contain again but one set. We need not now consider the details of the nuclear divisions in which this reduction occurs, but a brief comment on its essential character may nevertheless be helpful. The essence of meiosis is that it consists of two divisions of the nucleus accompanied by only one division of the chromosomes. During the first (prophase) stage of division contraction of the chromosomes begins as in mitosis, but the chromosomes are still undivided, and presumably in consequence of this they begin to come together in homologous pairs, a chromosome from one parent lying alongside the equivalent chromosome from the other parent. These chromosome pairs become shorter and thicker by their constituent paired chromosomes coiling about one another. Not until now does the delayed longitudinal splitting of the chromosomes occur. As a result of it

PROCESS	Pellia	Funaria	Nephrodium	ALTERNATION OF GENERATIONS
	Thallus bearing	Protonema and moss plant bearing	Prothallus bearing	gametophyte (haploid) x chromosomes
Gametes liberated and conjugate in water	archegonia and antheridia	archegonia and antheridia	archegonia and antheridia	conjugation
Embryo develop *in situ*	fertilised egg	fertilised egg	fertilised egg	
	sporogonium devoid of chlorophyll	sporogonium with photosynthetic tissues	free-living fern plant with stem, roots, and leaves, bearing sporangia	sporophyte (diploid) $2x$ chromosomes
Spores shed in air, and germinate in moisture	forming spores	forming spores	forming spores	reduction division (*meiosis*)

each chromosome pair is now represented by *four* daughter chromosomes (chromatids), lying alongside one another, but no sooner has splitting taken place than the chromatids begin to fall apart. It can, however, be seen in suitable microscopic preparations that the chromatids now hang together in certain places. This is because at these places, called *chiasmata*, the chromatids have broken and rejoined with new partners. It appears that without chiasma formation the mechanism of meiosis cannot exist. Following it pairs of chromatids withdraw along the lines of a spindle, as in mitosis, towards the two poles of the cell. At each pole there then follows a second nuclear division in which the units of each pair of chromatids separate from one another. Thus, as a result of the two divisions, both of which are essential, the chromosome number is halved. Both divisions are reductional, for at the first division the crossing-over of chromatids brings about segregation of some chromatid sections which were paired in the parental chromosomes, and this segregation is completed by the chromatid separation of the second division.

The recognition and interpretation of chiasmata represent a very great advance in the science of cyto-genetics. For detailed description and illustration of this modern theory of the mechanism of meiosis more advanced text-books of genetics must be consulted.

CHAPTER XIII

TISSUE ELEMENTS OF THE HIGHER PLANTS

In their organisation as what we may call "xerophyton" (the inhabitants of the dry land), the seed plants have come to display a high degree of differentiation, both of external organs and of internal structure. The external organs involve two extensive branch-systems, one below ground, the roots, and one above, the stem and leaves. Separate physiological rôles are associated with the separate organs, thus absorption of water and mineral salts with the roots, support and conduction of water and food with the stems, photosynthesis and evaporation with the leaves. In addition to this, leaves and stems and roots, particularly the two former, are often enormously modified, both structurally and in their metabolism, so that from them organs are produced capable of particular special functions. Thus swollen roots, or stems, or leaves may form food storage organs (as the carrot root, the underground stem of Iris, and the onion bulb respectively); stems may often form special perennating reproductive organs, both with food stores and with buds capable of growing into new plants (as the tubers of potato, crocus corms, and bulbs); leaves may become very greatly modified, to form the complex reproductive organs of the flower, which is the most intricate of reproductive structures in the plant world. These indicate only a very few, though perhaps the more important, of the modifications of form and function which the parts of the land plant shew. The study of the structural relationships of these modified organs (morphology) is a wide and interesting field of work which we cannot here enter. We shall seek instead, in the remaining chapters, to understand the main outlines of the physiology of the land plant, and in order to do this, we must make acquaintance with the simple structure of the root, stem and leaf organs in their normal unmodified rôle. We shall need, in doing this, to be familiar with the structure of the cells, the actual units out of which they are built, and as in the external organs of the land plant we have found division of labour and differentiation, so also we shall meet with it amongst these cell units (cf. also p. 56, Chapter IV). We shall find these cells aggregated into organised groups with definite features in common, and each such group or mass of cells is termed a "tissue." The feature common to the cells of a tissue may be structure (e.g. thin-walled ground tissue), or stage of development (e.g. adult or embryonic tissue), or function (e.g. conducting tissue). A large number of characters give names to tissues in this way, and naturally the same cells will often go into more classes than one. Thus fibres are found as part of a tissue called the xylem; they also constitute mechanical tissue, and may be part of the conducting tissue of the plant.

Simple Tissues. These usually consist of only one type of cell.

(1) *Parenchyma*. This is the simplest type of tissue and that from which all the other types can be derived either in the development of the individual or in the development of the race. The growing region of a stem or root is often called parenchyma; in it the cells are more or less *isodiametric*, they contain *living* and active *protoplasm* (with nucleus), they have thin *cellulose* walls, and retain their capacity for dividing (Fig. 6). As these cells at the growing point age, they give rise by differentiation to the other tissues of the stem, root and leaf, and so they may be said to be the basal type of which the others are modifications. Parenchyma often forms the living thin-walled "ground tissue" in which more complex tissues are formed, and it

is the tissue from which simple plants such as *Fucus* and *Funaria* are largely built up.

In the cells derived from the parenchyma of the growing region we shall find modifications especially in *cell-dimensions, cell-wall composition, cell-wall structure, the presence or absence of a protoplast* (i.e. living protoplasm).

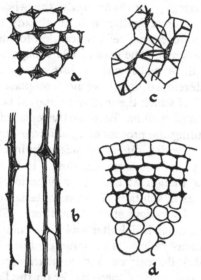

Fig. 48. *Collenchyma and stone-cells.* a, transverse section, and b, longitudinal section, through collenchyma cells from the stem of the potato plant, shewing the thick tapering strips of cellulose laid down on the cell-walls especially at the corners; d, transverse section through part of the cortex of a herbaceous stem shewing collenchyma especially developed as thickening on the tangential walls of the outer layers of cells; c, stone-cells as seen in either longitudinal or transverse section, shewing thick walls, small lumen and branched pits.

(2) *Collenchyma.* The cell elements of collenchyma have walls thickened with alternate layers of cellulose and pectin, but the thickening is not uniformly deposited on the thin primary wall, being laid down in longitudinal strips in the corners of the cells, and upon their tangential

walls. This gives the appearance indicated by the two sections (transverse and longitudinal) shewn in Fig. 48. Collenchyma occurs most commonly at the periphery of herbaceous stems such as the wallflower and potato, and its tensile strength and elasticity make it a mechanical factor of the greatest usefulness, for in conjunction with the turgor of all the living parenchyma cells, the whole stem may be maintained erect in the manner already described on p. 30. The collenchyma cells contain protoplasm.

(3) *Sclerenchyma.* The cells of sclerenchyma have essentially hard thick walls with a low water content, and when fully developed they have no protoplast inside them. The material of which the wall is composed is *lignin*. The original cellulose wall has been partly replaced and largely added to (during the process of *lignification*) by deposits of this substance, which differs considerably in chemical and mechanical properties from cellulose. It is much harder and stronger, though at least equally permeable to water and substances in solution. The two materials can be recognised readily by various staining reactions. Thus cellulose gives a purple colour in Schultze's solution (chlor-zinc iodide), and lignin stains yellow. The lignin contains both pentose (carbohydrate) derivatives, and aromatic (benzene ring) compounds, and both of these confer on the lignin definite staining properties. With phloroglucinol and strong hydrochloric acid it gives a very bright magenta-red colour, with aniline chloride or aniline sulphate it gives a bright golden yellow, and it readily takes up all kinds of aniline dyes. The contrast between cellulose and lignin may be very readily demonstrated by staining samples of good writing papers (made from plant tissue, like cotton or flax, entirely cellulose in nature) with samples of newspaper, brown paper, etc., made from wood pulp which contains a great deal of lignin. These staining tests should always be made upon tissues in examination of their constituent cell elements, but it

must be remembered that tissues may be unlignified during their early development which would shew lignin very strongly later on. Other tissues than sclerenchyma shew lignin, for example certain conducting cells (tracheids and vessels), but we may begin by considering the two types of sclerenchyma, stone-cells and fibres.

(*a*) *Stone-cells* (Fig. 48). These are isodiametric cells often lying in loose groups together, as in the tiny gritty specks in the flesh of the pear, or in continuous masses as in the hard shell of the coconut or hazel-nut. The walls are so heavily thickened that the space in the middle of the cell (the lumen) is reduced almost to nothing. Nevertheless beautifully branching *pits* through the thick walls shew the channels through which protoplasmic threads ran between the cells while they were developing. At maturity the cells have no living contents.

(*b*) *Fibres* (Fig. 49). These are elongate and pointed cells with walls so thickened that the lumen is quite small and the pits almost closed up. Although fibres are usually lignified, some are formed with cellulose walls even at maturity. The fibres occur together in long strands, their tapering ends closely fitting together. They so form a tissue of enormous tensile strength and of great elasticity; the strength of such vegetable fibres is in some cases as great as that of wrought iron, and the elastic properties are superior. Cellulose fibres, such as cotton and flax are widely used as textiles, and lignified fibres are the elements which confer properties of hardness and strength on such timbers as oak and ebony. The fibres at maturity are not living.

The Complex Tissues. There occur constantly in the stems, roots and leaves alike, of higher plants, two tissues called the *xylem* and the *phloem* (or wood and bast), which are always readily recognisable though not very closely answering the general description of a tissue as "a continuous organised mass of cells similar in origin and alike in form and general function." Each is composed of several distinctive elements, the xylem of *vessels, tracheids* and *fibres* as well as parenchyma, and the phloem of *sieve tubes* and *companion cells* as well as parenchyma and in some cases, fibres. These "tissues" are entitled to the term because, in spite of their complexity, they occur in definite positions in the plant organs, are constant in character, and are associated with quite definite functions, the xylem with water conduction, the phloem with the conduction of organic food in solution, and both to some extent with providing mechanical strength to the plant.

Xylem. (1) *Tracheids.* The fundamental unit of the xylem is the tracheid, which is the only element, save parenchyma, present in the wood of ancient fossil plants, and from which, in the course of evolution, both xylem vessels and xylem fibres have probably arisen. The tracheid is a single cell with lignified walls, and it has oblique or tapering ends; although the cell is fairly thick-walled the lumen is large and empty of contents. Each cell communicates with the next by bordered pits, the structure of which is illustrated in Fig. 49 b, c and c_1. The pits are places on the original thin primary wall upon which further thickening has not been laid down, and doubtless they facilitate the lateral conduction of water through masses of tracheids lying side by side. Very often local thickenings are deposited upon the walls of the tracheid cells in the form of horizontal rings (forming *annular tracheids*), or spirals (*spiral tracheids*), or parallel bars (forming *scalariform tracheids*), or in an irregular network (*reticulate tracheids*) (Figs. 49, 67; cf. figures of vessels, 51 and 66). In all cases, these thickenings lend strength to the cells and tend to prevent their collapse under pressure from neighbouring cells. Thus the wide lumen is kept open, and the tracheid maintains its double function of a sup-

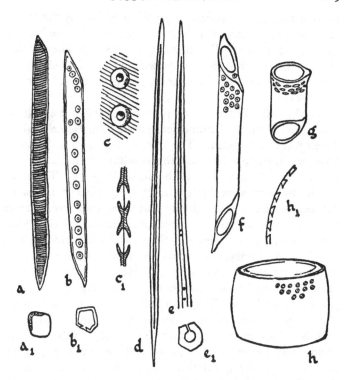

Fig. 49. Tissue elements of the wood. *a* and *b*, tracheids; *d* and *e*, fibres; *f*, *g* and *h*, vessel segments (the separate cells of which vessels are composed). The tracheids shew tapering walls, wide lumen and spiral thickening in *a* and bordered pits in *b*. *c*, bordered pits in surface view, and c_1 in section. The fibres can be regarded as tracheids modified in the direction of greater length, thicker walls, more tapering ends, smaller lumen and fewer pits. The vessel segments can be regarded as tracheids modified in the direction of shorter length, more obliquely transverse walls and wider lumen; *f*, *g* and *h* represent a series in such modification. (Pitting has only been shewn on part of these cells but it extends over all the walls.) a_1, b_1, e_1 and h_1 represent transverse sections through *a*, *b*, *e* and *h* respectively. (*d* to *h* modified from Eames and MacDaniels.)

porting and a water-conducting element. During the course
of evolution we may consider the tracheid to have become
specialised in two directions, the fibre, as already described, in the direction of greater mechanical efficiency,
and the vessel in the direction of greater water-conducting
efficiency.

(2) *Vessels.* The vessel is not a single cell, but a unit derived from a chain of cylindrical parenchymatic cells which
develop at the meristem. The intervening cross walls break
down at an early stage, and the resultant tube, rather like
a drain-pipe in structure, is the vessel (Fig. 50). The walls
round the very wide lumen become lignified in the same
manner as the tracheid walls, so that there are spiral, annular
and reticulate vessels, and bordered pits of similar kind
give communication between the vessels lying side by side
(Fig. 49). In transverse section, the vessels are conspicuous
for their almost circular outline, developed by pushing aside
other cells, and for the wide lumen and strong lignified wall
(Figs. 49 and 66). They probably constitute the most
efficient water-conducting channels of the land plant.

It is rather interesting that the small vessels and tracheids first formed and evident just behind the growing
point have chiefly spiral thickenings. As the cells around
them elongate (for they begin to differentiate even in the
elongating region of stem or root) the spirals are considerably drawn out, but they prevent for a long time absolute
collapse of the walls (Fig. 51). The other thickenings would
be essentially less satisfactory in this way and their absence
from this situation and their prevalence elsewhere is very
marked. In a so-called "mixed" xylem, we may attribute
support chiefly to the fibres and tracheids, conduction to
the tracheids and vessels, and such vital functions as there
are, to the parenchyma scattered about in it.

Phloem. *Sieve tubes and companion cells.* The paren-

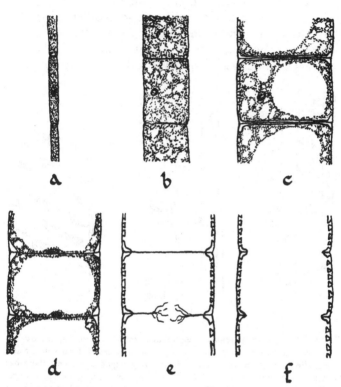

Fig. 50. The development of a xylem vessel. *a*, the vertical row of meristematic cells; *b*, the cells expanded; *c*, cells still larger and shewing thickening deposited on all the walls, even the transverse ones; shewing also pits through the thickening; *d*, protoplasm diminished in amount and the nucleus lying along the transverse cell-walls which are being dissolved; *e*, no living contents left in the cells and transverse walls breaking down; *f*, the mature xylem vessel composed of the empty, thick-walled segments. (After Eames and MacDaniels.)

chyma and fibres which occur in the phloem have already
been considered. They are found also in other tissues, but
the sieve tubes and companion cells are limited to the
phloem and are characteristic of it. The sieve tube, like
the vessel, is derived from a vertical row of superimposed

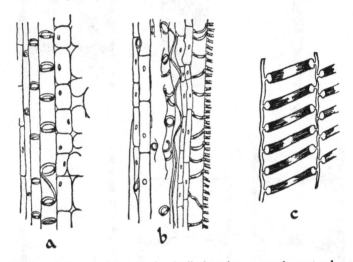

Fig. 51. Xylem vessels seen in longitudinal section. *a*, annular protoxylem
vessels as first differentiated; *b*, similar protoxylem vessels, both annular and
spiral, after undergoing extension in the growing region of a stem; there has
been complete rupture of the vessel walls in some cases; *c*, a mature spiral vessel
shewing the manner in which the spiral thickenings have been laid down upon
the pre-existing wall. (*a* and *b* after Eames and MacDaniels, *c* after Rothert.)

parenchymatic cells, but the intervening cross walls, though
much modified, do not break down completely. Instead,
they become perforated by numerous very fine pits through
which protoplasm passes, and even when cellulose is deposited on the walls, small scattered areas containing these
pits remain open over its whole surface. Such perforated
cross walls are called *sieve plates*, and they shew up clearly

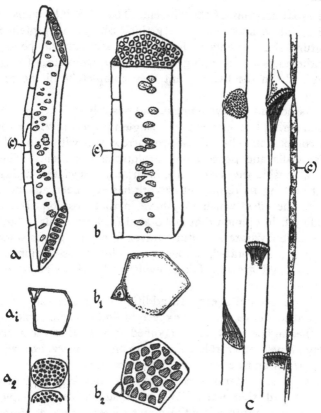

Fig. 52. Sieve tubes and companion cells. *a* and *b*, diagrams to shew the elements (single cells) out of which the sieve tubes are built; in *a*, the end walls are oblique and in *b*, horizontal, but in each case they carry sieve plates, *b* only one and *a* several. The sieve-tube elements are accompanied in each case by a file of narrow companion cells (*c*), and on the lateral walls shew rudimentary sieve plates. *a* and *b* are side views: a_1 and b_1 transverse sections; a_2 and b_2 surface views of the sieve plates; *c* is a longitudinal section through sieve tubes and companion cells as usually seen in stem or root sections; the two plates on the left have been turned over in cutting and shew their surfaces and the lower of the two shews protein strands pulled through the sieve perforations. The other plates shew the perforations and deeply staining slime strands below them. (*a* and *b* from Eames and MacDaniels.)

in cross sections of the phloem. The sieve tube walls are of cellulose and remain very thin. Slimy and protein in nature, the contents in longitudinal sections can be seen gathered as a "slime plug" below each sieve plate, although probably in the living plant they completely fill the cells (Fig. 52).

Associated in development, and possibly in function, with each sieve plate element is a single *companion cell*, or row of companion cells. These are narrow cells with thin cellulose walls, and dense living protoplasmic contents. They lie alongside the sieve tube, invariably associated with it, but nothing more than guesswork has ever been advanced, as to the effect which they have in that position. It is a widely-held opinion that elaborated food materials, protein and carbohydrate in nature, travel from the leaves where they are elaborated, through the phloem to regions of storage and growth. Direct evidence for this view is very scanty.

Secretory cells, etc. In addition to cells of the tissues described above, we may expect to find in the complexity of the higher plant, cells modified in yet other ways, and playing yet other rôles. Such are the *secretory cells*, with thin walls and dense protoplasmic contents, which secrete gum, oil or nectar, and *water storage cells*, with very large lumen and thin walls. We have sufficiently explained the main types to allow the direct investigation of simple plant structure, and to give some understanding of the part played by the tissues in the chief physiological processes. In this latter connection particularly, we should remember that the cells of almost all plant tissues are separated (or in some senses connected) by a system of *intercellular air-spaces*, which extends throughout the length of the plant body, and contains an internal gaseous atmosphere which is in direct contact with the outer air only through such limited areas as the pores of the epidermal stomata.

Table to shew the chief characters of the tissue elements.

Name of element	Occurrence in tissue	Thickenings on walls	Material of walls	Contents of adult cell	Dimensions of cell
(1) *parenchyma cells*	parenchyma (part of cortex, xylem and phloem)	thin, uniform	cellulose	living	isodiametric
(2) *collenchyma cells*	collenchyma (part of cortex)	thickenings in strips at corners	cellulose and pectin	living	slightly elongated
(3) *stone cells*	sclerenchyma (part of cortex)	uniform and *very heavy* save for branched pits	lignin	dead	isodiametric
(4) *fibres*	sclerenchyma (part of xylem, phloem, cortex)	uniform and *very heavy*	lignin or cellulose	dead	elongate and tapering
(5) *tracheids*	part of xylem	fairly heavy with extra annular, spiral thickenings etc. Bordered pits	lignin	dead	elongate, open lumen, oblique end walls
(6) *vessels*	part of xylem	fairly heavy with extra annular, spiral thickenings etc. Bordered pits	lignin	dead	long tube formed from a row of superimposed cell-units
(7) *sieve tubes*	part of phloem	thin, uniform	cellulose	dead	row of superimposed cell-units with perforate cross-walls
(8) *companion cells*	part of phloem	thin, uniform	cellulose	living	elongate

CHAPTER XIV

THE ROOT

As a rule the term "root" is limited to those parts of the plant body which develop below ground, but both stems and leaves frequently grow there also, and there are many examples of roots developing above ground. The typical root system produces at the end of each branch a definite growing region the chief characters of which have been already described (see pp. 65–76). Here, by the processes of rapid cell formation, extension and differentiation, all the mature tissues of the main axis are formed *behind* the *meristem* (region of cell division), as the root-apex is pushed through the soil. There is also a small but continuous production of cells by the meristem on the side towards which it is growing, and this forms the structure known as the *root-cap* (Fig. 53). The cells of the root-cap are loosely held together; they are continually rubbed off as the root-apex is pushed through the soil, and are as constantly renewed from behind. It is generally agreed that the root-cap serves to protect the delicate and very important meristematic tissues from injury by pressure of soil particles, and it achieves this no doubt, partly by forcing apart the soil grains mechanically, and partly by a kind of lubrication of the way by the mucilaginous contents of its broken cells. It is clearly indicated in the table on p. 75 that the dividing and elongating regions of the root, plus the root-cap, do not occupy, in the bean at least, more than the apical two or three centimetres. Such is the general case. Behind the elongating region the surface of the root becomes covered with a dense felt of thin white hairs, each a few millimetres long. This is called the *root-hair region* and it extends for a few centimetres only along the root (Fig. 54). Each single root-hair is a narrow, tubular outgrowth of the wall of an epidermal cell; the cell-wall is very delicate and translucent and is lined throughout its length with cytoplasm; a large vacuole extends throughout the cell, and the nucleus is usually present about the

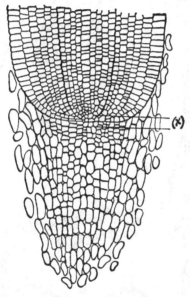

Fig. 53. The root-cap. Longitudinal section through the apex of the root of rye, shewing the structure of the root-cap and its origin from the meristematic region (*x*). Note the increase in dimensions of the cells (due to vacuolation), both behind the meristem and in the root-cap. (After Belzung.)

middle or end of the hair. If seedlings are grown in moist air, e.g. mustard grown on wet blotting paper, the root-hairs grow perfectly straight and at right angles to the root, but in soil they take on all sorts of distorted shapes according to the contact they make with the firm soil particles (Fig. 54). The outer layers of the cell-wall are mucilaginous and come

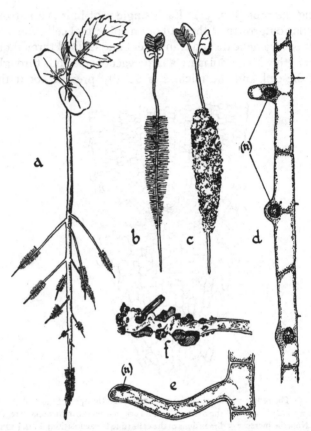

Fig. 54. Root-hairs. *a*, a hornbeam seedling shewing the limited region of root-hair production on the main and lateral roots. *b*, mustard seedling grown in moist air; *c*, mustard seedling grown in soil, shewing the soil particles adhering to the root hairs; *d*, epidermal cells of a root shewing the origin of root hairs as swellings from them; the nucleus (*n*) of the cell always occurs at the point of root-hair formation and in *e*, the adult root-hair, commonly occurs at the apex. The protoplasmic lining of the hair is indicated by shading. *f*, the end of a root-hair shewing close attachment to soil particles. (After Strasburger, Stevens, Sachs Haberlandt.)

into such close contact with the soil grains that the soil can
never be thoroughly removed from the root of a plant grown
in a garden soil save by breaking off the root-hairs. The life

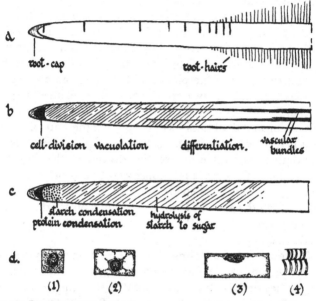

Fig. 55. Development at the root apex. *a, b, c* and *d* are a series of diagrams
illustrating different aspects of development in the same root (see p. 74).

a, external appearance shewing the extension of the root by the separation
of ten ink-marks originally made 1 millimetre apart, and also shewing the front
end of the root-hair region. *b*, regions recognisable microscopically as those of
cell-division, vacuolation and differentiation. *c*, regions recognisable by micro-
chemical examination. *d*, cells from the regions of cell-division (1), of early
vacuolation (2), of complete vacuolation (3), and of differentiation of protoxylem
elements (4).

of a root-hair is, in most plants, extremely short, so that the
root-hair region is only a few centimetres long and behind
it the root is again bare. The situation of the root-hairs so
close behind the elongating region, has some interest in that

the root-hairs form by their growth into the soil a kind of "foot-hold," from which the root apex, a centimetre or so away, is pushed through the soil by the elongating cells without buckling of the root.

From what has already been said, it will have become evident that the root-hair region corresponds in position with the region of differentiation of the tissues in the internal structure of the root. This is connected with the fact that the greatest part of the plant's water supply enters via the root-hairs. The newly formed conducting elements, spiral tracheids and vessels especially, occur in the root tissues in this region and doubtless serve to conduct the water upwards to the rest of the plant. Some few spiral elements, as mentioned earlier, are differentiated even in the region of elongation. This space relationship between the root-hairs and the newly formed conducting tissue can be readily seen by crushing the end of a thin root (e.g. of a mustard seedling) between two glass slides and examining under the microscope (Fig. 55).

Structure of the Young Root. A section across the root in the region just behind the root-hairs, shews that there all the tissues derived from the apical meristem (i.e. primary tissues), have become at least partly differentiated (Fig. 56). The cross section clearly shews a central *vascular cylinder*, surrounded by a *cortex* of thin-walled, closely packed parenchyma, and the whole is limited by a single layer of cells, the *piliferous layer*, which gives rise to the root-hairs. The vascular cylinder is separated from the cortex by a single layer of cells, the *endodermis*, and within the endodermis and separating it from the actual conducting tissues themselves, is a layer of parenchyma cells usually one cell thick, the *pericycle*. The xylem and phloem, which are the essential conducting elements of the vascular cylinder, occur in the root arranged upon alternating radial planes, in distinct strands separated from one another by parenchymatic ground tissue. The xylem strands seen in section are always wedge-shaped, with the apices of the wedges outwards (Figs. 56, 57). This is because the points of the wedges are formed by the *protoxylem*, that is the first xylem elements to be differentiated. These are the spiral tracheids formed in the elongating region of the root, and they are very narrow in diameter because thickened before much dis-

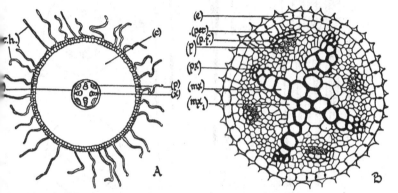

Fig. 56. Structure of the young root. *A*, diagram shewing production of root-hairs (*r.h.*), from the epidermal cells, and the immature vascular cylinder (*s*) containing four groups of xylem (*x*), alternating with four phloem groups (*p*); (*c*), cortex. *B*, large scale drawing of a similar vascular cylinder in the buttercup root. (*e*), endodermis; (*per*), pericycle; (*p*), phloem; (*p.f.*), phloem fibres; (*px*), protoxylem; (*mx*), metaxylem; (*mx₁*), metaxylem not yet lignified. (*A*, after Bower.)

tension of the cells had occurred. The next differentiated elements are called *metaxylem*, and they form to the *inside* of the protoxylem groups. They consist of pitted vessels of which those towards the central axis of the root have an increasingly large diameter, for they represent elements thickened at later and later times, and consequently from cells further and further distended. In slender roots the meta-xylem of the separate xylem wedges meets in a solid mass

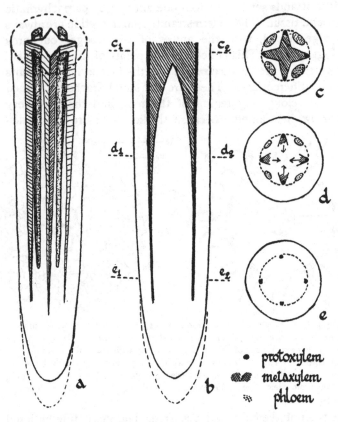

Fig. 57. Differentiation of vascular tissue behind the root apex. *a*, the root apex as it would appear if all the tissues were made transparent save the phloem (dotted) and the xylem (shaded). *b*, longitudinal section of the same root; *c*, *d* and *e*, transverse sections of the same root taken at the levels $c_1 c_2$, $d_1 d_2$, $e_1 e_2$, respectively in figure *b*. All the diagrams shew the centripetal development of the metaxylem (see arrows in *d*) and the exarch position of the protoxylem, the first formed vascular elements.

in the middle, but in large roots a *pith* of parenchyma remains. The sequence of development of the xylem from without inwards is called *centripetal* development (as contrasted with the stem which has *centrifugal* development), and the protoxylems are said to be *exarch* (as in the stem they are *endarch*). The exact nature of the sequence of development can be more easily made out by reference to Fig. 57. As a rule the sieve tubes and companion cells of the phloem strands begin to differentiate later than the xylem, so that they are seen to be immature in sections cut just behind the root-hair region, although they can be recognised because they occur as definite tissue areas in the appropriate positions. In many roots, at a little later stage strands of fibres develop within the phloem, and lend to it properties of tensile strength, which are important because of the strains coming upon the root in its "anchorage" of the plant in the soil. The number of xylem wedges (or phloem groups) in a root is roughly constant for any given species of plant, thus the broad bean usually has four.

The *endodermis* is a structure of considerable interest, although its significance in the life of the plant is by no means agreed upon. It forms a cylinder, a kind of chimney, of uniformly brick-shaped cells, completely enclosing the vascular cylinder of the root. Although the walls are initially made of cellulose, early in development fatty substances (*cutin*) are deposited in continuous bands within the substance of all the lateral and transverse walls (Fig. 58). When seen cut across in longitudinal or transverse section, the endodermis shews this band as small dots on the common walls between its constituent cells, and if the cells are plasmolysed, it will be readily seen that the cytoplasm, though shrinking from the tangential walls, is firmly attached to the cutin strips in the other walls, and still stretches unbroken across the cell in the tangential plane. As this is

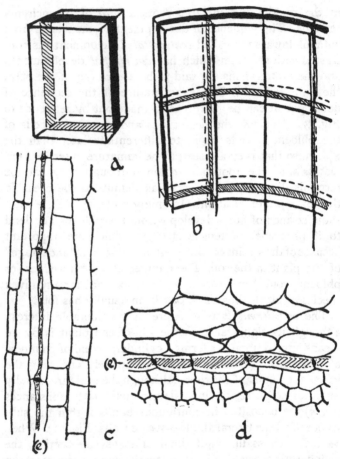

Fig. 58. The endodermis. *a*, diagrammatic drawing of an isolated endodermal cell shewing the band of fats deposited in the radial walls; *b*, diagram to shew the disposition of a few of the endodermal cells as arranged in a cylinder in the root; *c*, the endodermis as seen in the root in longitudinal section; *d*, the endodermis as seen in the root in transverse section. Here, as in *c*, the cytoplasm has been plasmolysed and has contracted from the cellulose cell-walls save at the regions of fat deposition. (*c* and *d*, after Eames and MacDaniels.)

true of each cell, and as the cutin in the walls prevents the diffusion of water or water soluble substances along them, the endodermis forms an uninterrupted barrier between the cortex and the vascular cylinder. All solutions and substances passing between the two must therefore pass through the living protoplasm of the endodermal cells, and be subject to what we may roughly speak of as its "control"; not a conscious control of course, but an influence certainly. Given a mechanism of this kind, we may easily imagine that it plays a considerable part in the process of translocation of organic and inorganic substances about the root. Various theories as to its action have indeed been put forward, but there is no general agreement about them. We may only note its occurrence as a "physiological barrier," and tentatively use it in explanations of such frequent phenomena as the occurrence of abundant starch within the cortex whilst it is absent in pericycle and vascular cylinder, and so forth.

The *pericycle* is not composed of cells of particular structure, but is a tissue which is named chiefly by its position between the endodermis and the conducting tissues, and which may include both parenchyma and sclerenchyma.

The Adult Root. *Branching.* Behind the root-hair region the root-hairs wither and fall off, and then from the primary root, branch roots (secondary roots) appear. These arise in a quite definite manner from *within* the root, and push their way out from the vascular cylinder to the outside. Usually one lateral root arises in the pericycle outside each protoxylem group, so that in a tetrarch root (a root with four xylem groups), like the bean, the lateral roots arise in fours, in a triarch root in threes etc., and the lateral roots also appear on the outside of the root arranged in a number of corresponding vertical rows; this is very evident in such storage roots as the parsnip and radish, and in water roots like those of the willow (Fig. 59). The cells of the pericycle begin to divide and form a small protrusion which makes its way, partly by mechanical force and partly by the secretion of

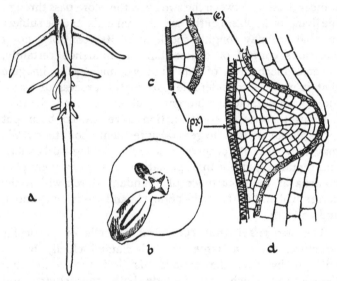

Fig. 59. The origin of lateral roots. *a*, bean root shewing lateral roots arising in four vertical rows which correspond with four xylem groups in the stele. Note the split in the cortex through which the lateral roots emerge. *b*, cross section of a young root shewing the origin of two lateral roots from the pericycle region and each behind a xylem group. The larger root has almost broken through the cortex. *c* and *d*, longitudinal sections through a part of a young root shewing the origin of the lateral root as a meristem just within the endo-dermis (*e*) and to the outside of the protoxylem vessels (*px*).

enzymes which break down the cells, through the cortex to the outside of the root. By this time a definite root-cap and apical meristem have been formed, and differentiation begins of the cells between the young root and the vascular tissues of the main axis, so that in time they come into connection.

The ruptured piliferous layer of the main root makes no connection with that of the young root protruding through it. According to conditions some of the laterals stay small and others develop rapidly, and new ones are continually produced by the main axis and by the laterals themselves, so that a branching system is formed which comes in contact with a very large volume of soil.

Secondary thickening. In a number of plants, such as the grasses and lilies among others, no tissues are formed in the roots save from cells produced at the apical meristems. This naturally limits the thickness of the adult root to a few millimetres, though the length may be very great. In a very large number of plants however, both roots and stems augment the original primary axes by continual increase in thickness, so that mature forest trees may have roots several decimetres across, organs equivalent in size to the main branches. Such continuous increase in size is brought about by a process known as secondary thickening, which is too complex to be considered in detail here. Briefly, we may say that there is present in any such thickening root, a cylinder of meristematic cells which is continually active in the production of new cells. These differentiate into xylem and phloem, and so continually add to the bulk of these tissues. Such a meristem is called a "cambium," and almost the whole of the bulk of any woody stem can be attributed to its activities. The tissue formed in greatest amount is xylem, so that the mass of tissue within a big root or stem is so-called "secondary" wood or "secondary" xylem. When the outer part of the stem of a tree is peeled off, the natural break takes place in the fragile cells of the cambium; the piece broken off is commonly called bark (though not in an exact botanical sense) and the whole of the rest is xylem tissue composed of wood-tracheids, fibres, vessels, and parenchyma. The phloem of the tree is present

in the broken-off "bark." The same case holds closely for
the thickened root (Fig. 62, *c*), and though we cannot now
consider the origin of these tissues we may note that they
are made essentially of xylem and phloem, which play the
same conducting rôle here as in the simple unthickened
root which we have already described.

The physiological rôle of the Root. A land plant such
as an oak tree growing in ordinary soil, can be resolved into
two branching systems, one aerial and one subterranean,
connected by a main axis which is partly above and partly
below ground. The aerial system consists of stems which
repeatedly branch and occupy the space of a large hemisphere, at the surface of which the finer branches end in a
dense canopy of leaves. The subterranean system occupies
an equally large hemisphere of soil, and the finest roots are
to be found especially round the outside of it. These fine
roots, clothed as they are with root-hairs, present a very
large surface in close contact with the soil, and in the same
way the flat plate-like leaves borne by the stem system
present a great surface in contact with the air. This large
surface makes for great efficiency in absorption of water
from the soil in the one case, and for great efficiency in
evaporation of water into the air in the other case. If the
connecting system between the absorptive and evaporating
systems is capable of permitting the flow through it of a
rapid water current without undue resistance, it will be clear
that such a plant growing in moist soil will constitute a
path by which large quantities of water will escape from
the soil into the drier air. This stream of water through
the plant is called the transpiration stream and it is present
in all land plants. Some idea of the rate of the process can
be obtained from the calculation that in an eighteen weeks
growing season a single sunflower plant lost six gallons of
water, a great many times the bulk of the plant itself.

The biological significance of the transpiration stream to the plant, lies less in actual movement of *water* out of the soil, than in the movement of substances which are dissolved in it. A plant which lives, like *Chlamydomonas* or *Volvox*, free swimming in a stream or pond, is constantly bathed in a dilute solution of inorganic salts, all to some degree or other ionised. Many of these, together with carbon dioxide which is present in the same form, are essential to the growth of the plant. The movement of the water round the plant brings fresh supplies to the surface of the cells as constantly as ions are absorbed, but with the land plant growing in ordinary soil, the question of supply is more difficult though the need for the mineral elements is the same. In the absence of water movement, the requisite salts present in very dilute solution would reach the root-hairs only by diffusion, and the nearby soil being exhausted, further supplies would reach them extremely slowly. It is now generally held that water moves freely through the soil only in the rare condition where free water exists in the spaces between the soil particles, and that absorption of water by the root-hairs cannot in fact give rise, under ordinary conditions, to a current fed from a large volume of soil. In most soils it is probable that there is an enormous production of fine short-lived roots, which constantly grow into fresh parts of soil, extracting from them both water and dissolved substances. Large quantities of mineral elements are in fact absorbed, and the transpiration stream inside the plant further serves to carry these substances to the leaves and growing points above ground, in which places they undergo synthetic processes which end in their incorporation in the plant body.

In the most general sense we may analogise the transpiring plant with the simple apparatus shewn in Fig. 60, and we may expect the physical conditions affecting the one to affect the other similarly. The two porous pots represent the evaporating and absorbing systems, and the glass tube the conducting system. Undoubtedly transpiration normally proceeds in the plant in this manner, and we may say that evaporation at the leaf surfaces is the motive force

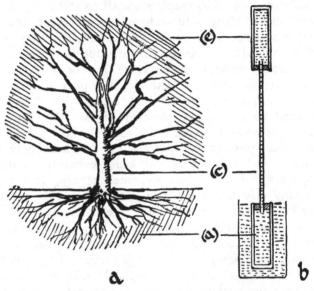

Fig. 60. The transpiration stream. *b*, two porous pots connected by a glass tube and filled with water, the lower in a vessel of water and the upper in dry air; by evaporation from the upper pot water constantly passes through the system. The tree, *a*, has similar regions of absorption (*a*), conduction (*c*), and evaporation (*e*), and though much more complex, behaves in a roughly analogous way. (See text.)

of the whole process, "pulling up" columns of water through the xylem from the roots and through the root-hairs from the soil. In some circumstances however, the movement of a water stream through the plant must go on by other means, for plants will often give off water into absolutely saturated air, and many plants, such as the vine in the spring, will force up long columns of water from their roots when the stem above is cut off and no leaves are yet developed. These cases do not however affect our ideas on the significance of the transpiration stream, or on its usual mechanism.

Soil conditions. It is quite clear from what we have said, that the condition of the soil and the relation of the roots to it will have a great effect in determining the rate at which water uptake can go on. It will be not so much the actual amount of water present in the soil, as the forces tending to retain the water in it against the pull of the roots, which will affect the plant. The soil water is classified into fractions of different availability to the plant roots. The water which sinks under the influence of gravity and drains out of a soil sample, is called *gravitational water*; another fraction remains held in the soil crevices between the soil particles and stretching as thin films over them, this is the *capillary water*; lastly there is *hygroscopic water*, which is retained when the capillary water has been removed; it is water absorbed in the soil colloids and combined with soil silicates. The gravitational water is as readily available as free water, the capillary water is held by surface tension forces of two or three atmospheres, and the hygroscopic water is held by forces which may be as much as a thousand atmospheres. There is no sharp line of division between the given soil water fractions, and as a given soil dries up, first the gravitational, then the capillary, and then the hygroscopic fraction diminishes, and all the time the forces tending to keep the water in the soil increase, and render absorption more difficult. Thus the degree of dryness of the soil will have a great effect on the rate of water uptake by the root. The *kind* of soil will also have a great effect; thus a plant might be able to take water from a sandy soil containing 10 per cent. of water, whilst quite unable on account of the finer particles and greater surface forces involved, to obtain any from a clay soil containing the same percentage of water. The kind of soil will also affect the total amount of available water that the soil can hold, and so will affect the plant by the length of time its water content will last the transpiring plant.

Other factors such as temperature and osmotic concentration of dissolved salts may also come into play; thus the absorption from the soil will diminish as the temperature falls and will cease as the soil water freezes. Round salt marshes and salt lakes, e.g. Utah and the Lake Chad region in the Sudan, the high salt content of the soil causes very high osmotic pressures in the soil solution. The effect of this on water absorption, together with the toxicity of such strong salt solutions, probably accounts for the fact that only a very few highly specialised plants can exist in these places.

Under the soil conditions generally found in Great Britain the capillary water is the fraction most made use of by the plant. The close application and the growth of the root-hairs round the soil particles, puts their cell-walls into the closest contact with the thin water films; in the absence of such contact the utilisation of this water fraction would be greatly retarded if not wholly prevented. It is worth noting that with capillary water present in the soil, large air spaces coexist, from which all the root tissues obtain the oxygen necessary for their respiratory processes. In water-logged soils the air spaces are full of gravitational water, oxygen is no longer available to the roots, water absorption ceases, and the roots die.

Thus a good soil should be well aerated and well watered, and should allow easy penetration of the plant roots, in addition to containing in the soil solution a range of chemical substances necessary to plant growth.

Salt intake. For healthy growth of the plant to take place, the soil solution must contain substances which yield the seven following elements: nitrogen, phosphorus, sulphur, magnesium, potassium, calcium and iron. These are present in the soil as salts, but in the soil solution these are ionised to a very considerable extent, and the essential elements are absorbed as ions. This is readily seen from the fact that of two ions derived from a single salt, one is often absorbed much more than the other. Similarly all the various ions in solution are absorbed at different rates, so that if the root of a growing plant is placed in a culture solution of suitable salts in known proportions, after a short time the proportions of the elements in the solutions will have been considerably changed. This is spoken of as differential absorption. The uptake of the ions by the root-hairs is probably not connected with the inflow of water except in so far as this brings new supplies of ions to the surface of the root-hair. The actual process of the taking in of the ions seems to depend on their continual utilisation, or "fixation" inside the root-hair cells. In this way they are removed from solution; the concentration of them inside the cell diminishes, and more diffuse in. Naturally in this the permeability of the protoplasmic cell lining to the various ions plays a part, and this in fact varies for different ions.

Thus it will be clear that the mineral substances taken in by a plant do not enter with the transpiration stream, carried in with it by mass flow, but are separately taken in, each to a different extent, by a process of diffusion which is closely affected by the condition of the living protoplasm of the root-hair cells.

Direction of growth. It has been indicated earlier that the direction of growth of the root is determined by a series of tropisms. Thus the main root is sensitive to gravity, and

responds to the gravitational stimulus by turning downwards. It has been shewn that the region which "perceives"
the gravitational stimulus is situated largely or entirely in
the apical 2 millimetres. The response itself is made in the
elongating region of the root, so that *conduction* is necessary
between the region of perception and response as in animal

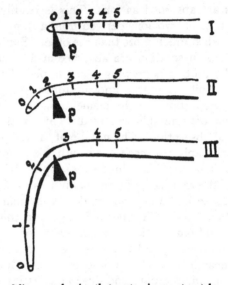

Fig. 61. Series of diagrams shewing that geotropic curvature takes place in the
extending portion of a root. *I*, the root placed horizontally and marked into
equal divisions behind the apex; *II* and *III*, stages in the subsequent geotropic
curvature; only in the segments 0 to 3 which are those which have expanded
most has curvature taken place. *p* is a fixed index point. (After Sachs.)

responses, although no corresponding nerve structures have
ever been found in the plant. Recent research work makes
it increasingly probable that the stimulus resulting in plant
tropisms is conducted by the movement of definite chemical
substances which are comparable with hormones in the
animal body, and which move backwards from the region

of perception and modify the growth rates of different parts of the region of response. Recent research has permitted some of these growth-controlling substances to be isolated, identified, and even synthesised. One of the best known is β indolyl-acetic acid. Commercial preparations of these substances are now on the horticultural market, where they are of importance chiefly on account of their powers of stimulating the formation of new roots in cuttings. In the case of a geotropically stimulated root the response takes place by a decreased rate of cell extension on the downward side of any root tilted out of the vertical, and this *growth* reaction brings about downward curvature. Since such curvature is limited to the growth region of a root, a curve once made is not straightened out again, for the growth region is always passing on towards the root apex, and any attempts to correct one curve will merely give a second curve below it in the opposite direction (Fig. 61, and also Fig. 67). The main root alone is positively geotropic, the lateral roots from it grow at an angle, more or less constant, to the vertical, and the laterals from these are not sensitive to gravity. Plant roots are sensitive to certain degrees of moisture, and in a dry soil grow towards sources of water. The mustard seedling is an example of the rare cases of plants with roots which turn away from the light. All these tropisms, positive geotropism, positive hydrotropism, and negative phototropism, would tend to send the root down into the soil; but there are other tropisms which also affect its course. Thus in badly aerated soils roots will often grow towards higher concentrations of oxygen, and slight wounding leads to a curvature of the root tip away from the source of the stimulus. All these responses in the heterogeneous soil environment, together with varying degrees of root-development produced at different seasons and in different soil layers, give to the root system the complex and irregular appearance with which we are familiar.

CHAPTER XV

THE STEM

The term "stem" is usually given to that part of the axis of the plant which grows into the air and bears the leaves and reproductive organs Both in internal and external structure the stem and root are distinguishable.

General structure. The stem always terminates in a *bud*, which is a particular type of apical meristem in which the main region of dividing cells consists of a small conical zone at the stem apex which constantly produces small lateral meristems all round it. These develop continuously into lateral organs, especially leaves, which in a more or less immature condition enfold the small central meristem. Thus the stem apex might be said to have a kind of "multiple" meristem, in contrast with the simple meristem of the root. The lateral organs produced by the stem clearly have an origin quite different from those of the lateral roots, since they arise from surface bulges in the apical meristem, and not from pericycle tissues within the adult axis. As the cells produced by the meristem pass through the phase of elongation, the lateral outgrowths are stretched further and further apart, until they shew the arrangement typical of the adult stem. The leaves may then be seen to occur singly, in pairs, or in whorls (rings) of three or more, at definite points on the stem. These points of leaf insertion are called the "nodes," and the stem regions between them are called "internodes." On the adult stem the leaves are borne in a remarkably regular and ordered way, thus in the dead-nettle family they occur in pairs alternately set, each pair at right angles to the pair above; in other families they are spaced out exactly in a spiral manner, and in a few they occur in repeated whorls. It has been suggested that these exact spacing systems are very close to those which would expose the foliage leaves to the light with least overlapping, whilst preserving a uniformly balanced plant body throughout development. Cases of this kind of actual approximation to some mechanical principle are often found in plant construction, and we may suppose their present occurrence to be due to elimination of the less satisfactory, and preservation of the more suitable in the severe competition of natural selection, continued over very long periods. On the other hand, the regular leaf arrangement may be the expression of an inherent tendency in plant tissues due to the interaction of the cells upon each other, or to something of that kind.

The leaf differs from the other lateral outgrowths of the stem (lateral branches), in the fact that it generally bears in the axil between leaf stalk and the stem from which it grows, a small bud (an axillary bud), which later develops into a lateral branch. Thus we may say that the lateral organs originate in pairs in the meristem, each small bulge of meristem which represents a leaf having a smaller lump of meristem just above it representing an embryo bud. As a general rule the leaves are flat, thin, green structures, and stems are elongate and branching and bear foliage leaves. But in some cases these structures may be modified as we have earlier suggested (p. 190), and then the nature of any lateral appendage can be determined by finding whether it occupies the position of a leaf from another lateral structure in its axil, or whether it occupies the position of a stem, itself occurring in the axil or some other structure. For example the prickles on a gorse bush may by this method be readily resolved into stem spines and into leaf spines, and the validity of the conclusion can be tested by examining the seedling gorse plant which possesses quite normal foliage

leaves and shoots, and shews stages in their transition to
spines.

Anatomy. In the apical part of a young stem, which has
not yet begun a process of secondary increase in thickness
like that in the root, we find in cross section that, as in the
root, there is a natural separation into epidermis, cortex and
vascular cylinder. The *epidermis* resembles that which will
be described in the leaf, and it is similarly pierced, though
at wider intervals, by stomata identical with the leaf stomata.
The nature of the *cortex* depends closely on the nature of
the stem; it is often loose parenchyma as in the root, but it is
often collenchymatous towards the outside, and in such green
stems as the gorse it closely resembles the palisade tissue of
the leaf. In such cases it is easy to regard the stem cortex
as a downward prolongation of the tissues of the leaf stalks
over the surface of the vascular cylinder. Grounds have been
found for the wide application of this view, in what is called
the "leaf-skin theory" of the stem structure. The innermost
layer of the cortex forms the *endodermis*, which has the
structure of the same tissue in the root, save that it tends to
be less continuous and to have less uniformity in the presence
of fat deposits in the walls. It is often referred to as the
"starch sheath," because it contains conspicuously large and
abundant starch grains.

The vascular cylinder which occupies the centre of the
stem is usually much wider than that of the root, and differs
considerably from it in structure. It is made up of a large
number of bundles arranged in a ring round a central core
of parenchyma called the *pith*, which may in some cases
break down to leave a hollow space. Each bundle consists
of xylem on the inside and phloem on the outside, both
arranged, it should be noted, on the *same* radius and not, as
in the root, upon alternate radii. In a large number of stems
the *vascular bundles* are, from the beginning, coalescent to

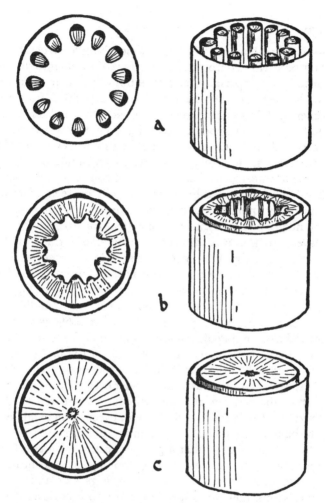

Fig. 62. Stem structure; diagrams to shew the disposition of xylem and phloem in three types of stems; phloem black, xylem shaded. The left-hand series of drawings shews the appearance in transverse section and the right-hand series a view of a cut portion as it would appear if all save the vascular tissue were removed from the stem. *a*, a ring of isolated bundles each with xylem inside and phloem outside; *b*, the xylem and phloem form a continuous cylinder from which the protoxylem groups project into the central pith. *c*, reduced in scale much more than *a* and *b*; a stem (or root) in which secondary thickening has produced an almost solid cylinder of xylem surrounded by a thin layer of phloem. Stems of types *a* and *b* when young, may later develop the structure of *c*.

form a continuous cylinder, to which bundle structure can only be attributed from the projections of xylem into the pith. Such stems especially merit the term "vascular cylinder"; in the rest the vascular bundles are separated from each other by fairly wide bands of parenchyma, and so remain quite distinct in appearance. They do not, however, pursue an entirely independent course in the stem even then. If an adult leafy stem of this kind is soaked in eau-de-javelle and so made translucent (i.e. cleared) the course of the vascular bundles will be readily visible. It will be seen that they repeatedly branch and fuse with each other, especially at the nodes, where vascular strands (leaf-traces) go out to the leaves. In this way the separate bundles form a kind of cylindrical network of vascular tissue in the stem which is thus provided with a path of translocation of water and food materials from one side of the plant to the other (Fig. 63). The condition of a cylinder of separate vascular bundles may persist in quite large plants such as the sunflower, but in most the beginning of secondary thickening from a cambium, soon welds both xylem and phloem into unbroken tissue cylinders such as those in the old tree root or stem (Fig. 62 *c*). All the tissues of the vascular cylinder between the outermost part of the phloem and the endodermis, are regarded as *pericycle*, a tissue which includes both parenchyma and fibres. The bundle structure of the xylem and phloem often extends into the pericycle, so that a group of *pericycle fibres* is found opposite each phloem group, forming a discontinuous cylinder of fibres outside the phloem (Figs. 63 *b* and 64). In some stems, however, especially those with coalescent vascular bundles, the pericycle fibres may also develop as a continuous cylinder.

A transverse section of a single vascular bundle will shew that the xylem and phloem tissues of it are of the same composition as the same tissues in the root; the phloem

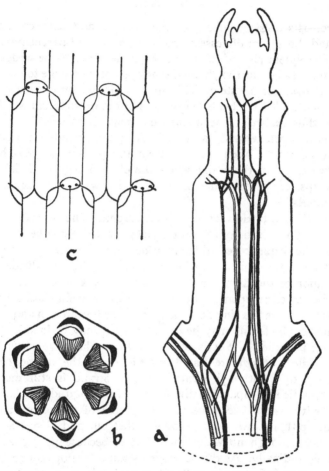

Fig. 63. The course of the vascular bundles in the stem of Clematis. *a*, diagram of the stem as it would appear if all the tissues save the vascular strands were made transparent; note the branching and fusing of these strands at the places where leaves come off (nodes). *b*, transverse section through an internode to shew six separate bundles with xylem (shaded), phloem (white), and pericycle fibres (black). *c*, diagram of the vascular system opened out into a flat sheet; four leaf-bases are shewn with three vascular strands passing into each. (*a*, after Nägeli.)

consists of sieve tubes, companion cells and parenchyma, and the xylem of vessels, tracheids, fibres and parenchyma. The xylem part of the primary bundle tends to be wedge-shaped as in the root, but with the apex of the wedge towards the *centre* of the stem. This apex is formed by the narrow thick-walled tracheids and vessels of the protoxylem which is thus *endarch*. The metaxylem consists of much larger, more open vessels, arranged in quite definite straight rows, which often fan out from the protoxylem to the broad end of the wedge where the phloem begins. Usually strips of parenchyma separate the rows of metaxylem vessels.

Differentiation at the Stem Apex. The key to the endarch position of the protoxylem is naturally to be found in the sequence of differentiation of the xylem elements behind the apical growing region. Behind the dividing region of the stem, the cells of the cortex and pith may be seen to enter upon the phase of extension, whilst still separating them is a ring of tissue with dense protoplasm and big nuclei. This ring may be complete or it may be in strands, but in either case it is the tissue from which the vascular bundle is later differentiated and it is termed the "desmogen" (Fig. 65). Although each desmogen strand is still meristematic it is capable of division only in a longitudinal plane. As it exists between the extending cells of the cortex and the pith, its own cells become pulled out and elongate in shape, though no wider than before, because they divide longitudinally. As the desmogen strand develops (or as we examine it further from the stem apex), in it may be seen the enlarging cell elements which will later form the vessels. Close to the inner side of the desmogen strand, some of the elongate cells begin to shew spiral thickening upon the walls. Very soon these become lignified and the protoplasm dies in the small protoxylem tracheids and protoxylem vessels

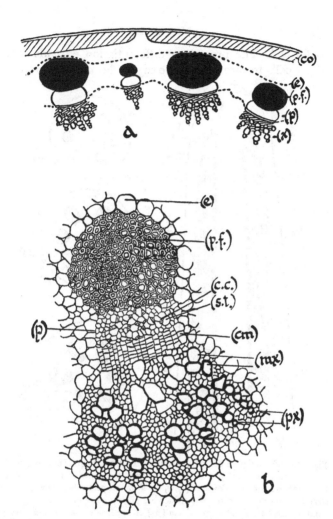

Fig. 64. Vascular structure of the Sunflower stem. *a*, diagram of transverse section of part of the stem shewing four vascular bundles with xylem (*x*), phloem (*p*), phloem fibres (*p.f.*), endodermis (*e*) and collenchyma (*co*). *b*, large scale drawing of a single bundle shewing sieve tubes (*s.t.*) and companion cells (*c.c.*) in the phloem, metaxylem (*mx*) and protoxylem (*px*). Between xylem and phloem is the meristematic zone [cambium, (*cm*)] which adds secondary xylem and phloem to the bundle as the stem gets older. (*a*, after Thoday.)

so formed. The process of differentiation extends to the
larger cells towards the inside of the desmogen strand, and

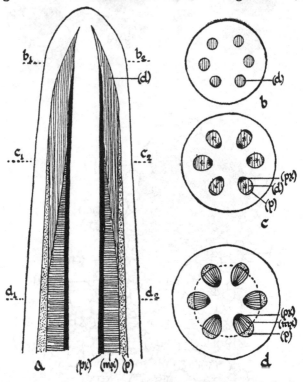

Fig. 65. Differentiation behind the stem apex. *a* is a diagram of a longitudinal
section through a stem apex and *b*, *c* and *d* are of transverse sections cut through
the same stem at the levels $b_1 b_2$, $c_1 c_2$, $d_1 d_2$, respectively. The desmogen
strands (*d*) of meristematic tissue which are present close to the actual growing
point differentiate first to protoxylem (*px*) on the inside, and then phloem (*p*)
forms on the outside; metaxylem forms centrifugally (see arrows in *c*) until the
bundle differentiation is complete. The protoxylems thus remain endarch.

from them the metaxylem vessels and tracheids become
differentiated (Fig. 66). The protoxylem elements shew up

with great clearness in stained longitudinal sections through
growing buds, e.g. brussels sprout. The embryonic leaves so
thickly cover the stem apex that many desmogen strands in

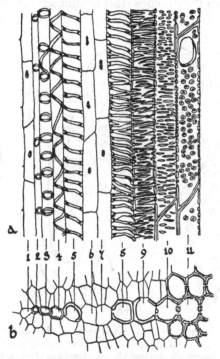

Fig. 66. Structure of the protoxylem and metaxylem; part of a vascular bundle
seen *a*, in longitudinal, and *b*, in transverse section. 2 to 5 protoxylem elements
thickened either by rings or spirals; 1, 6 and 7 parenchyma; 8 to 11 metaxylem
vessels thickened in various ways, 8 scalariform, 9 and 10 reticulate, 11 pitted.
(After Eames and MacDaniels.)

their earliest state seem as much related to the leaf meristems
as to the meristem of the main stem. Such a strand as one
of these first shews differentiation of protoxylem at a point
opposite the leaf base although the desmogen strand can be

seen extending up the leaf and down the stem axis. The differentiation of protoxylem progresses from this point both up and down the strand, so that continuous protoxylem soon exists between the leaf and stem (Fig. 67). The differentiation

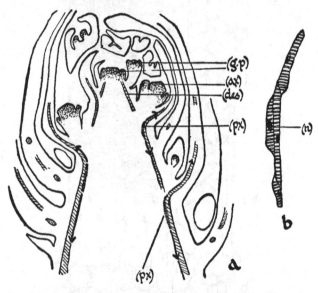

Fig. 67. Differentiation behind the stem apex. *a*, diagrammatic longitudinal section through the stem apex of the brussels sprout shewing growing points (*g.p.*) dotted; desmogen strands (*des*) shaded; on the inner side of the desmogen strands, protoxylem elements (*px*) are being differentiated, and this differentiation, beginning at a leaf-base, is shewn by arrows to be extending up into the leaf and down the stem, eventually to join all the leaf and stem bundles into a continuous system. (*ax*) axillary bud. *b*, spiral tracheids from the very young protoxylem; the cells still contain protoplasm and nuclei (*n*).

of the desmogen strand proceeds down the stem until the new vascular strand fuses with others in the stem to form part of the vascular cylinder. Differentiation of the meta-xylem follows suit and so complete vascular connection is

established between leaf and stem. It follows naturally that
in the *leaf-trace* so formed the xylem should be on the upper
side, and the phloem on the lower side, a condition to be
met with also in the vascular structure of the veins of the
leaf.

The axillary bud in the leaf axil may develop at once, or
may lie dormant for a longer or shorter time, but if such
buds grow immediately they can be seen to produce small
desmogen strands themselves, and by differentiation in the
cortical cells intervening, they come to establish complete
vascular continuity with the main vascular cylinder.

Within the desmogen strand the phloem begins to differentiate into sieve tubes and companion cells somewhat later
than the protoxylem, and at the opposite side of the strand.
Outside the desmogen strands the pericycle fibres become
evident early, though their thickening and lignification may
continue for a long time. The cuticle, a thin varnish-like
layer over the epidermis grows thicker, probably by continuous exudation of fatty materials over the surface, where
they are rendered more permanent by oxidation; the epidermis differentiates into stomata and brick-shaped epidermal
cells, and the cortical cells assume their characteristic mature
structure, whether collenchyma, parenchyma or photosynthetic tissue. At the same time all the tissues of the leaves
differentiate and develop their mature form. So all the
primary tissues of the stem are formed, yet within each
desmogen strand there usually persists, between the xylem
and the phloem, a strip of tissue which remains meristematic.
This is the cambium, from which proceeds the secondary
growth in thickness which is responsible for the great bulk
of xylem and phloem tissues in practically all trees and in
many herbaceous plants.

The transition between the vascular structure of the
primary stem and primary root takes place fairly abruptly

in an intermediate region about soil level, where the xylem and phloem bundles change their disposition in the axis, the protoxylems change their position with regard to the metaxylem, and all the smaller structural changes occur.

The mechanical rôle of the Stem. We have suggested that the disposition of tissue in the root system is related to the tensile strains involved in anchorage in the soil, and in a similar way we may explain the arrangement of tissues in the stem in terms of a mechanical supporting function. The more or less solid cylinder of wood forming the bulk of a tree trunk is mechanically a pillar almost solidly made of strong mechanical tissue, capable of supporting the heavy load of the tree crown and of withstanding the heavy stresses induced by it. In herbaceous plants other arrangements are found. Thus, as we have mentioned earlier (p. 30), many stems are maintained erect by the turgor of living paren-chyma cells within a closed cylinder of some much stronger tissue such as collenchyma. Rigidity is also given to such stems by the lignified elements of the wood, especially the tracheids and fibres, and above all by the strands of pericycle or of cortical fibres. The disposition of these extra-vascular fibre-strands often coincides with accepted mechanical prin-ciples. Thus they very frequently are disposed in a cylinder round the periphery of the stem, which is the most eco-nomical use of material if a structure capable of withstanding bending strains from all sides, is to be constructed. In many stems the fibres are disposed in the form of **I** girders or **T** girders, both common engineering units, and these are particularly common also in the leaf. The sight of a large herbaceous plant with flowers and foliage borne on a relatively slender stem and unbroken in a high wind, brings some realisation of the mechanical efficiency of the system, and the use of the xylem as timber, and the fibres as textile threads, testifies to the strength of the material employed.

Conduction in the Xylem and Phloem. Besides supporting the foliage leaves and flowers and fruits in the air, the stem system must supply them with considerable amounts of water from the roots, and it must be a means of conducting soluble organic food from the leaves where it is synthesised, to storage organs and growing points both above and below ground. We have so far assumed that the two conducting functions are peculiar to the xylem and phloem respectively. It is interesting to note the type of evidence on which this assumption rests.

(*a*) *Evidence that the xylem is the path of water-conduction.* The form of the xylem units, and the correlation of their abundance in various regions with the rate of water movement through these regions to different parts of the plant, afford *a priori* reason for supposing that they are the path of the transpiration stream, but other experimental evidence has been forthcoming. Thus if cut shoots of plants are allowed to absorb solutions of dyes such as eosin, the dye will be found to have risen through the stem and to have stained only the xylem elements. These dyes can be shewn to stain all other tissues if brought in contact with them, so that the experiment seems fairly valid. If the wood at the cut end of a woody shoot is laid bare and dipped into melted fat, so that all the vessels and tracheids are occluded by it, whilst the phloem is left open, it will be found that water absorption is prevented, when the shoot is put back into water. Further, the bared wood of any tree, e.g. a foot length of a twig stripped of all soft tissues, will shew rapid conduction of water through it under quite small pressures. Accumulated evidence of this kind has left no doubt that the xylem *is* the path of the transpiration stream in the plant, but opinion still differs as to the mechanism by which the flow is brought about. A diminishing number of people believe that it is primarily due to the *vital* activities of the living parenchyma cells of the stem, in a rather unspecified kind of pumping action. A great number of plant physiologists hold mechanistic views of the nature we have already set out, i.e. that it is largely due to evaporation from the leaf surface. It was for a long time said that this explanation was impossible, on the grounds that the large tensions in them would break the water columns, and gas bubbles would form at a certain height in the tracheids and vessels, so that the water would not rise above such a point. Recently it has been shewn that water columns as narrow as those in the vessels possess such great tensile strength that even containing air bubbles they can exceed in length the great height of such trees as the eucalyptus and the red-wood (*Sequoia*).

(*b*) *The rôle of the phloem.* Although there is fairly general agreement about the xylem, the theory that the phloem is the path of the conduction of the organic food material is still largely unproved. No particular fitness for this rôle is apparent in the structure of the sieve tubes or companion cells, and experimental evidence is very scanty. Such as there is consists in ringing experiments. In these the phloem is cut away in a ring, sometimes complete and sometimes not, extending all round a woody stem. The xylem is left intact and then the interference produced in the movement of food materials is measured in some way or other. When carried out on fruit trees this ringing usually results in great swelling of the cortex and phloem tissues just above the cut, as if food were accumulating there. Although this is suggestive, more precise measurement of the accumulation of organic food has been lacking until recently, and now it is only possible to say that modern research tends to justify the old hypothesis that the phloem *is* the path of conduction of organic matter about the plant. In a general way we may speak of a water conduction *up* the plant in the xylem, and of a conduction of organic food *down* the plant in the phloem, though doubtless to some extent reverse currents occur in both.

Tropisms. The tropisms of the stem differ from those of the root in the type of reaction, and type of stimulus which produces them, rather than in the way in which the tropism is brought about. The stem is negatively geotropic and grows away from the centre of the earth,—a plant kept in the dark so as to exclude the influence of light will, when placed horizontally, soon shew an upward curvature of the stem. As in the root, this takes place in the apical elongating region, and is due to a changed growth rate on the two sides of the stem. Therefore, as also in the root, a curvature once made is not straightened out (Fig. 68), and no curvature takes place save at stem apices. The same does not quite hold for all stems; a certain number like the grasses retain the power of growth in the tissues surrounding each node. If such a stem as this is placed horizontal, rapid growth goes on at the lower side of the nodes, so that at each node some curvature ensues, and the whole stem curves upwards although the internodes remain quite straight. In this way young grass stems recover far more readily from being trodden down or knocked over than do ordinary stems.

Stems are also very noticeably positively phototropic, growing towards the source of illumination, and though under normal conditions both light and gravitation produce upward growth, where they come into competition the phototropism is usually dominant, and stems may even be made to grow vertically downwards by illumination from below. The commonest phototropic response of the stem is, however, the lateral bending towards the light, which is most commonly seen in plants grown in windows etc., and which is due, like other tropisms, to the unequal growth rates of the two sides of the growing region. As with the root, the

region of response is not the region of perception, a fact
which can be easily demonstrated. A pot of *Setaria* (Italian
millet) seedlings 2 or 3 cms. high is placed in lateral illumination, the apical 5 millimetres or so of some of the shoots
having been covered, each with a tiny cap of tin-foil. All the
uncovered shoots will be found to bend sharply to the light
about 1 cm. from the apex, but the capped shoots, though

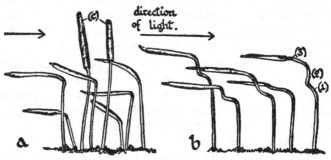

Fig. 68. Experiment to shew phototropism of grass seedlings. Seedlings of
Setaria were grown in complete darkness, and some of them were capped with
small cylinders of tin-foil (*c*). At this stage all were erect. They were then exposed to lateral illumination from a window. The same evening they had assumed
the positions shewn in *a*; the uncovered ones had bent over, but those in which
the sensitive apical region was darkened had remained upright. The seedlings
were left overnight and *b* shews the appearance of some of them next day. Note
the double curvature of the axis shewing phototropic response on the first day, (1);
vertical growth in response to geotropism during the night, (2); phototropic
response on the second, (3).

uncovered in the responsive region will remain upright
(Fig. 68). As is in the case of geotropic response of the root,
the stimulus conduction is due to the backwards movement
from the apex of special growth-regulating substances. It
has been shewn that when the extreme apices of shoots are
cut off and placed upon agar jelly the growth substances
diffuse into it. Blocks of the jelly then put upon the cut
ends of the decapitated shoots cause increased rates of

extension. This shews that extension can be controlled by
readily diffusible substances produced by the stem apex.
If such a block containing growth substance is placed
excentrically on the cut end of a decapitated shoot, then the
side below the block grows faster than the other, and a
curvature results. Further experiments suggest that the
effect of lighting from one side is to cause a *redistribution*
of the growth substance produced by the tip, more of it
going towards the darkened side, and there causing greater
extension. Very much more, however, is yet to be learnt of
the mechanism of response to stimuli, and it is now one of
the most active fields of physiological research.

Etiolation. Light is responsible in the plant for other
effects than phototropisms. Varying light intensities produce
various structural changes during the development of stems
and leaves, which are not by any means fully understood.
Thus the form and tissue composition of leaves and stems
grown in sun and shade respectively may differ very greatly,
and plants grown in complete darkness have a very characteristic appearance. They are said to be *etiolated.* The stems
are very long and straggling, with no lateral branches and
leaves which remain very small; it is as though growth in
length had replaced the development of lateral organs. The
tissues are poorly differentiated and the stems are very weak,
but perhaps the most striking character is that leaves and
stems are quite yellow owing to the non-formation of
chlorophyll and the accumulation of a yellow intermediate
compound. The rapid elongation of the etiolated stem has
often been related to the increased growth of the darkened
stem in the phototropic response, but it is clear that in
continued darkness other factors of considerable complexity
must come in.

CHAPTER XVI

THE LEAF

The leaf is an organ which probably shews a greater range of modification in form and structure than either stems or roots, but we have space here only to consider the nature of the ordinary green foliage leaf. This typically has a broad thin *blade* or *lamina*, supported on a *leaf-stalk* or *petiole*. Often at the base of the petiole two small wing-like out-growths called *stipules* are to be seen, e.g. in the pea, bean and rose leaves, but in many plants they are inconspicuous, temporary or absent. From the petiole, the vascular strands of the leaf-trace branch into the whole of the vein system of the leaf, affording it at once a supporting skeleton and a means of translocation of water and food materials. In a certain number of plants such as the grasses and lilies (which we have already mentioned as having no secondary increase in thickness), a number of equal main veins traverse the leaf longitudinally parallel to each other, and finer veins connect them (Fig. 69). The more usual type of *venation* is, however, a system of net-veining in which a main vein gives rise to laterals of a smaller size, and these to other laterals, and so on, until the whole leaf is supplied by the finest end veins. In both parallel and net-veined leaves the fusions of the branching veins with each other cut up the leaf tissue into separate areas and the tiniest of these areas, usually only one or two square millimetres in area, are called vein-islets. The skeletons of rotted leaves gathered from any woodside ditch will shew the beautiful and intricate pattern of the venation with great clearness, and it may even be seen that each vein-islet is supplied by a tiny, blindly-ending vein. In such a manner the whole of the green tissue of the leaf is in closest contact with the vascular supply.

As the leaf is in direct continuity with the stem axis so the veins are the prolongation of the vascular cylinder of

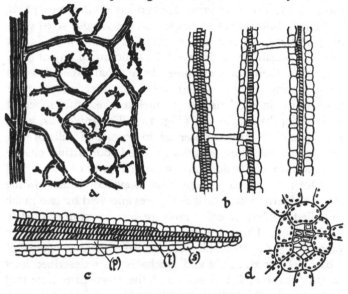

Fig. 69. Vascular system of the leaf. *a*, surface view of part of a net-veined leaf to shew the branching and fusing veinlets; each black line represents a single tracheid. *b*, similar view of a parallel-veined leaf shewing the spiral tracheids enclosed in a sheath of parenchyma cells. *c*, longitudinal section through a bundle-end shewing the sheath (*s*), the spiral tracheids (*t*) extending to the end of the bundle, and the phloem (*p*) extending not so far. *d*, transverse section through a veinlet shewing tracheids and phloem in section surrounded by a sheath of parenchyma cells containing chloroplasts. (After Sachs, Stevens, and Eames and MacDaniels.)

the stem, the epidermis which covers both surfaces of the leaf is in continuity with that of the stem, and the living green tissues of the leaf (the mesophyll) represent the stem cortex.

Anatomy. A cross section through an ordinary green foliage leaf shews, between the limiting layers of the upper and lower epidermis, the *mesophyll* divided into *palisade tissue* towards the upper surface, and *spongy mesophyll* towards the lower surface. Embedded in the mesophyll, and cut fortuitously at all angles, we see various sized veins in transverse, oblique and longitudinal section. It is not however possible to understand the structure of any of these tissues, in their solid or three-dimensional aspect, from a cross section alone, so that we must also consult their appearance in sections cut parallel to the leaf surface at various depths in the leaf (Fig. 70). The *epidermis*, when seen in surface view, is observed to consist of cells of very varying shape, often irregularly lobed, but fitting closely together like a mosaic to give a surface quite uninterrupted save for the *stomata*. These, more prevalent usually in the lower epidermis, are small oval pores enclosed by two much modified epidermal cells, more or less sausage shaped in surface view. These are the *guard-cells*, and by their turgor movements the stomatal pore between them is opened or closed. Their structure and mechanism are described later in this chapter; it will suffice at the moment to note that they contain chloroplasts which are absent from the other epidermal cells and when the epidermis is stained in iodine they often may be seen to contain abundant dark starch grains, whilst the rest of the epidermis is quite colourless. In the transverse section of the leaf the epidermal cells always shew a rectangular shape, which indicates that they are flat (or tabular) cells, with a lobed outline forming a kind of mosaic pavement on each surface of the leaf. The cell contents are living cytoplasm, nucleus and vacuole. The cell-walls are usually of cellulose and are often thickened on the outside, but greater protection than these can afford is given to the leaf by a layer of *cutin*, spread like a coat of varnish over the outer surfaces of both upper and lower epidermis. This layer is the *cuticle* and it is almost completely waterproof and gasproof. Its continuity is broken however at the stomata. In some plants the deposition of cutin goes on even in the walls of the epidermal cells, which by this process become much stronger and more durable. The presence of the cutin can be readily demonstrated by the use of some stain such as Sudan III which gives fats a bright red colour.

The palisade tissue consists of one or two layers of cylindrical green cells set below the upper surface of the leaf and at right angles to it. In surface sections these cells appear circular, and they touch each other only here and there, since numerous intercellular spaces occur among them. These are far less obvious in transverse sections of the leaf, and may be quite overlooked in thick sections. In transverse sections of the leaf the palisade cells are elongate and rectangular. Each is vacuolate, with living contents and thin cellulose wall, but its most striking feature is the dense crowding of disc-like chloroplasts inside it. They occur here in numbers unmatched in the cells of any other part of the plant, and they constitute the chief photosynthetic organs of the green leaf and therefore of the plant.

The spongy mesophyll or spongy parenchyma lies between the palisade and the lower epidermis. In this the parenchyma cells are less elongate than those of the palisade and contain far fewer chloroplasts. In addition they are irregularly grouped in branching chains, forming a loose spongy tissue which encloses enormous air-spaces. The volume of the air-spaces in the spongy mesophyll must often be much larger than the volume taken up by the cells themselves. Stomata are usually abundant in the lower epidermis, and always round each of them a particularly large "sub-stomatal" air-space is present in the spongy mesophyll. As in the palisade,

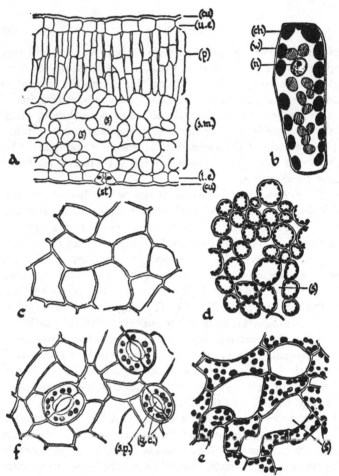

Fig. 70. Leaf structure. *a*, transverse section through the blade of a cherry-laurel leaf shewing cuticle (*cu*), upper epidermis (*u.e.*), lower epidermis (*l.e.*), with a stoma (*st*), palisade tissue (*p*), penetrated by narrow air-spaces, spongy mesophyll (*s.m.*), with abundant large air-spaces (*s*). No cell contents have been shewn in the figure. The stoma shewn is an old one and mesophyll cells have blocked the air-space behind it. *b*, side view of a single palisade cell shewing the cell-wall (*w*), nucleus (*n*), and chloroplasts (*ch*). *c* to *f*, sections cut *parallel* to the leaf-surface; *c*, upper epidermis devoid of chloroplasts and with no stomata; *d*, palisade cells shewing circular section, chloroplasts lining vertical walls, and narrow air-spaces (*s*); *e*, spongy mesophyll shewing irregularly shaped cells with scattered chloroplasts and wide air-spaces; *f*, lower epidermis shewing three stomata, in each of which two guard-cells (*g.c.*) containing chloroplasts enclose the stomatal pore (*s.p.*); the other epidermal cells are devoid of chloroplasts.

the cells of this tissue are alive and have thin cellulose
walls.

The vein structure is also extremely interesting. For the
most part the veins occur in the spongy mesophyll, the
larger veins forming distinct ribs which bulge out on the
lower surface of the leaf. As mentioned earlier, the xylem
and phloem of the veins is so disposed that in cross section
of the leaf the xylem lies uppermost and the phloem below.
The constituent elements are those found in the stem and
root, and doubtless the xylem elements give considerable
strength to the leaf skeleton. The netted skeleton of veins,
together with the natural turgor of the leaf cells, serves to
maintain the leaf rigid and exposed to the light. The xylem
elements in the veins are generally supplemented by other
mechanical tissues. Especially common is a tissue of fibres
forming a complete cylinder round the vascular strand and
thicker on the upper and lower sides of the bundle. Examples
exist of all stages between this and the very common type in
which a bundle of fibres occurs on the upper and lower sides
of each bundle. As of the stem, a great deal has been written
to shew with what economy of space and material these
mechanical tissues have been disposed. In a number of
leaves cellulose strengthening in the form of collenchyma
is laid down in the midrib, but none of these strengthening
tissues continues to the smaller leaf-veins. In these, mechanical support to the mesophyll is of far less importance
than a contact between conducting elements and parenchyma
cells which would be interfered with by thick-walled fibres
or collenchyma. The finest veins of all gradually taper
away and end in the vein-islets. The conducting elements
become reduced in number and the xylem consists of a few
spiral or pitted tracheids only, and may end in a single one.
The phloem ends before the xylem, and the whole bundle
end is completely surrounded by a continuous sheath of

parenchyma through which must go on all material exchanges between the conducting tissue and the mesophyll
(Fig. 69).

The Leaf in relation to photosynthesis. Since the cuticle
is almost impervious to gases, it follows that practically all
the carbon dioxide used by the leaf in photosynthesis must
enter from the air via the stomata. From the stomata it
diffuses through the substomatal air-space (Fig. 71) and the
smaller air-spaces running continuously about the leaf, and
it goes into solution in the water held in the moist cellulose
walls of the mesophyll. Thence by hydro-diffusion it reaches
the chloroplast surface, where, in the presence of light, it
undergoes photo-reduction.

If we consider an illuminated leaf, then within the active
chloroplast surface the concentration of carbon dioxide
approximates to zero, the concentration just outside the
chloroplast surface will be higher, in the air-spaces of the
leaf higher still, and in the outer air it will remain about 035
per cent. The rate at which carbon dioxide flows into the
leaf will depend upon (1), the activity of the chloroplast,
which will in turn be controlled by other factors such as
the intensity and colour of the light, the temperature, etc.
When the chloroplasts are active the concentration of carbon
dioxide at the outer chloroplast surface will tend to be very
low, and carbon dioxide will diffuse from the outer air to
the chloroplast surface on account of the difference in concentration (i.e. diffusion potential) maintained between the
two places. When the chloroplast is less active, the carbon
dioxide tends to reach higher concentrations at its surface,
and the diffusion gradient from the outer air becomes less
steep (i.e. the diffusion potential is less), and so carbon dioxide
diffuses into the leaf more slowly. It will be evident that
the rate of diffusion to the chloroplast surface must be
controlled not only by the activity of the chloroplast working

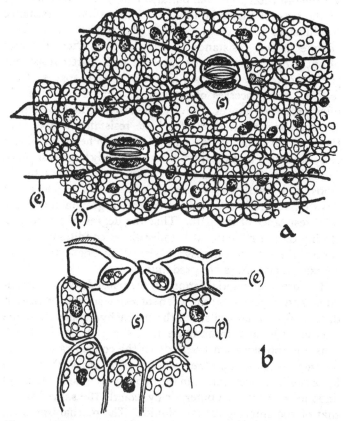

Fig. 71. The leaf. *a*, strip from the surface of an Iris leaf seen from above, the walls of the large empty epidermal cells (*e*) are seen overlying the closely packed palisade cells (*p*), each of which contains a nucleus and numerous chloroplasts. Two stomata interrupt the continuity of the epidermis and below each there is a large air-space (*s*) in the palisade. *b*, part of a similar leaf seen in cross section so as to shew the two guard-cells of the stoma and the substomatal air-space.

via the diffusion potential, but also by (2), all the characters of the path of the carbon dioxide which afford a resistance to diffusion. Even when the chloroplast activity is constant, change in this resistance to diffusion will affect the rate of entry of carbon dioxide into the leaf. Little resistance is offered to free gaseous diffusion in the air-spaces of the leaf, but the processes of solution and hydro-diffusion will offer larger resistances. Nevertheless so long as the stomata are open and themselves offer little resistance to gas flow, the total resistance of the path is small. When the stomata close, however, they create such an extremely large resistance to diffusion that the carbon dioxide current practically stops, and when they are nearly shut it will be clear that the stomatal resistance will be far the largest factor in the total resistance of the path. Thus the degree of opening or closing of the stomata will under these conditions largely control the rate of photosynthesis. With open stomata of course, other factors are predominant.

The organic material formed in the palisade cells accumulates, and as the concentration of each substance rises it diffuses from these cells into the parenchyma sheath of the bundle ends, and from these into the phloem, in which constant translocation keeps down the concentration. The oxygen, which is the other product of photosynthesis, also increases in concentration in the photosynthesising cells, and finds its way into the outer air by exactly the same path as that of the entering carbon dioxide. The mechanism of all these movements seems to be diffusion from places of higher to places of lower concentration.

The Leaf in relation to transpiration. The water lost in transpiration by the leaf, enters it by the xylem elements of the veins, and via the sheath of parenchyma round the bundle-ends it passes into the mesophyll cells, and extends through their permeable cellulose walls. From these moist

surfaces exposed to the intercellular spaces evaporation goes
on, and the water vapour accumulating in the internal atmosphere of the leaf escapes by diffusion through the stomata
into drier air outside the leaf (Fig. 72). So long as the
concentration of water vapour in the air is small enough,
there will be a gradient of vapour pressure from the vascular

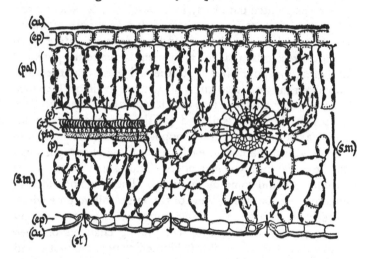

Fig. 72. Diagrammatic transverse section through a typical leaf shewing
cuticle (*cu*), epidermis (*ep*), palisade cells (*pal*), spongy mesophyll (*s.m.*) and
stomata (*st*). Two veins are seen traversing the spongy mesophyll, the one on
the right cut transversely and the other cut almost longitudinally; both shew
xylem elements (*x*), phloem (*ph*) and parenchyma sheath (*p*). The arrows
indicate the way in which water may pass from the veins to all the cells of the
leaf and how after evaporating from the cell-walls into the intercellular air-spaces
it diffuses through the stomata into the outer air. (After Stevens.)

bundles to the external air, and water will continue to diffuse
as vapour from the leaf. The rate of such vapour movement
will be dependent, as in the carbon dioxide movement, on
resistance to the flow, as well as on the potential (the difference in vapour pressures). The path and its resistances are

substantially those of the carbon dioxide current, so that we may safely conclude that the stomata, when closed or nearly closed, have a great influence on the rate of transpiration. When they are open other factors play a larger part in controlling its rate, for example, air dryness and the availability of the soil water.

As was pointed out briefly in Chapter XIV, the whole plant, root, stem and leaves can be regarded as a path of water-conduction between the soil and the drier air. Here again we must recognise that the movement is due to the potential of a greater vapour pressure of the soil water than the water vapour in the air. Two kinds of path are available by which the soil water can move into the air; (1) through the capillary spaces of the upper soil, and (2) through the roots, stems and leaves of plants. Each opposes certain resistances to this flow, and the flow in each case is inversely proportional to the resistance which it offers. Thus some water moves by one path and some by the other. The plant provides an effective mechanism in view of the large soil volume which it reaches. Large absorbing surfaces, large evaporating surfaces, and effective water-conducting channels between the two, are the secret of this high efficiency, the first found in finely branched roots and finer root-hairs, the second in the wide flat leaves and their large internal evaporating surfaces, and the third in the efficiency of the xylem elements. Doubtless the transpiration stream *is* to some degree useful to the plant in ways we have already considered, but given the existing plant structure it *must* flow through the plant whether useful or no.

Measurements of the transpiration rate. If a potted plant is arranged with the whole of the pot enclosed in a water-proof bag sealed closely round the plant stem, any loss in weight of the whole apparatus can be taken to be due to water loss from the exposed leaves and stem of the plant, since the changes in weight by photosynthesis and respiration are of a relatively negligible order. By successive weighings the course of transpiration may be measured, and the effect of external conditions upon the rate may be investigated. It can be shewn, for instance, that transpiration is more rapid in dry than in moist air, in warm than in cold air, and in air currents than in still air.

Other experiments involve the use of a cut twig, the cut end of which is sealed in a water vessel of small volume, open only by a capillary tube to a vessel of water in the outer air. The plant water-intake, which is closely similar under constant conditions to the water loss, is then measured by noting the rate at which an air bubble is drawn along the capillary tube in the water stream flowing along it. This and similar experiments involve rather unnatural conditions for the experimental plants, potting on the one hand, and detachment on the other. It is extremely difficult to investigate the transpiration rate under absolutely natural conditions, and probably the former of the two methods mentioned is the more suitable.

The Leaf in relation to respiration. The oxygen consumed in the respiration of the mesophyll cells of the leaf at night may be supposed to enter from the outer air via the stomata[1] and air-spaces, just like the carbon dioxide used in photosynthesis. It is probable, however, that the high oxygen concentrations accumulated in cell sap and air-spaces of the leaf during the day, serve as considerable reserves available for respiration at night. Certainly such a process is of greater importance to submerged aquatic flowering plants to which oxygen is less readily available from the surrounding water, than it is to land plants from the air.

[1] The stomata are seldom completely shut even at night and the oxygen concentration in the air is so big, and respiration uses so little oxygen, that the process is probably not affected by ordinary stomatal closure.

We may expect also that of the carbon dioxide formed by respiration at night much never escapes into the outer air, but, accumulating in the leaf, is utilised in photosynthesis on the following morning.

Stomatal movements. Several different types of stomata occur in land plants, but each is made of two more or less sausage-shaped guard-cells, which are so thickened that when they become turgid they swell apart by pressing into the neighbouring epidermal cells. In this state they shew a stomatal pore between them. As they become less turgid the pore grows smaller, and when they are flaccid the pore becomes closed by the apposition of the inner walls of the two cells. It is clear that two parts are needed in such a mechanism, (a) a means of changing the cell turgor, (b) a means whereby the cell turgor change distorts the shape of the guard-cell. The second of these is due to a differential thickening of the wall of the guard-cell. In one common type of stoma, each guard-cell has two heavy strips of thickening on the inside (concave) walls of each of the sausage-shaped cells, while the outer (convex) walls remain thin (Fig. 73). Increased turgor of such a cell causes expansion of the thin wall, and the thicker wall acting as a hinge, the cell becomes more curved than before. Since the two cells are joined at their ends this curvature opens the pore between them. When there is loss of turgor in the guard-cells the pressure of the surrounding cells and the elasticity of the cell-walls diminish the curvature again and close the pore. Other stomatal types differ from this in the disposition of the thickening material, but the principle on which they work remains the same.

The basis of the turgor change in the guard-cells is often, as in the elongating stem or root, to be found in the sugar-starch balance (see p. 74). In other cases it may be due to turgor changes of the whole leaf, such as take place in

wilting. In the former case although chloroplasts occur in
the guard-cells, and not in the other epidermal cells, and
although they doubtless produce, in the first place, the

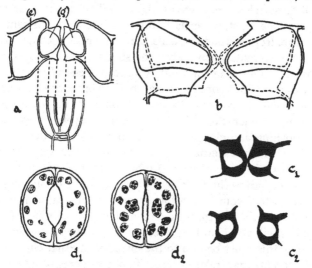

Fig. 73. *Stomata*. *a*, transverse section and surface view of half a typical stoma,
the two diagrams connected by dotted lines to shew the correspondence between
them. *g*, guard-cells; *e*, epidermal-cells. *b*, diagram shewing in dotted lines the
position of the guard-cells when flaccid, and in continuous lines their position
when turgid. c_1 and c_2, drawings to the same scale of a stoma of another type
in the closed and open condition; in the latter the lumen of the guard-cells has
assumed a much more nearly spherical shape. d_1 and d_2 shew in surface view
the stomata of Verbena at 9 a.m. and 6 p.m. respectively. In d_1 the chloroplasts
in the guard-cells contain little starch, the cells are turgid and the stoma is
wide open; in d_2 large starch grains have been formed in the chloroplasts, the cells
are flaccid and the stomatal pore is almost completely closed. (*a* and *b* after
Schwendener, *c* and *d* modified from Lloyd.)

materials for this mechanism, it is not to the photosynthetic
capacity of the plastids but to their starch-condensing
and starch-hydrolysing powers that the turgor change
is due. The ordinary palisade cells of the leaf accumulate

starch by day, and form sugar from it at night, but in the guard-cells the critical concentration of sugar varies in such a way that sugar accumulates by day and rapid conversion of it to starch takes place at night. Where high sugar concentrations exist, water is "drawn in" by the high osmotic pressure from surrounding cells, the guard-cells become turgid and the stomata open. Thus the stomata open by day and close by night (Fig. 74), and it appears generally true that guard-cells which contain no starch when open, shew granules of it when closed, and those with some starch when open, shew more when closed (Fig. 73). In this way we can see that the opening and closing of the stomata is controlled by the way in which external conditions affect the sugar-starch reactions in the guard-cells. Light has been shewn to favour sugar formation (high critical concentrations of sugar) and darkness to favour starch formation (low critical concentrations), but the rhythmical day and night opening of the stomata will also continue in leaves cut and kept illuminated, so that this must be due to the effect of internal change on the starch-sugar relationship. The sequence of chemical change which causes alterations by day and night, or in light and darkness, in the sugar-starch balance, is very little understood.

When the leaf wilts, i.e. when its general turgor and rigidity fail through water losses too rapid to be made up from the root, the loss of turgor affects the epidermal cells before the guard-cells. Consequently the turgor of the latter asserts itself further by bulging into the epidermal cells, and the stomata open wide. This is only short-lived however, for soon even the guard-cells lose water to the air and to the rest of the leaf and become flaccid and then the stomata close.

Significance of stomatal movement to the plant. We have indicated that stomatal movement is produced by light and darkness, by a day and night rhythm, and by acute

water loss. It may also be caused by mechanical shocks such
as shaking. The significance of such movement to the plant
used in the past to be a very thorny debating ground. In
particular the stomata were supposed to be a kind of water
conserving device, closing at times of severe water strain,
but we have seen that they do not work until the leaf is
already wilted.[1] It is more probable that their opening by
day and in the light is related to the photosynthetic process,
but the significance of closure at all seems rather vague and
unrecognisable. The greatest value of stomatal movement
to the plant would seem to consist in closing at night when

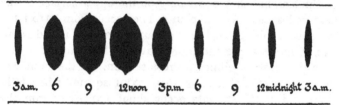

Fig. 74. Diagram to shew the daily rhythm of opening and closing of the stomata.
The black area represents the stomatal pore of the leaves of a Verbena plant in
July in Arizona. (From data given by Lloyd.)

photosynthesis is no longer possible, so preventing water loss
at this time, but even so the lowered night temperatures cut
down transpiration so greatly that this seems a small gain.
Possibly the mechanism may have behaved in a manner
more obviously advantageous to the plant, in some ancestors
of present day plants under other conditions, for it is difficult
to attribute *no* useful significance at all to a mechanism
occurring so widely and in such diverse plants. It is possible,
however, to regard the movement as incidental to the structure of the stomata, and as unavoidable as the transpiration
stream, and like it only accidentally useful. At most it may

[1] By a secondary effect on the sugar-starch balance prolonged wilting may
even bring about re-opening of the stomata.

be said that the movements tend to limit transpiration to those periods, more or less, when photosynthesis is possible.

Significance of the presence of stomata. The importance of the *presence* of the stomata, as distinct from their movements, is more evident. Their presence throughout an epidermis which is gas-proofed with cuticle, permits access of carbon dioxide for photosynthesis to underlying tissues, and yet prevents to a large degree desiccation of the leaf. The size, abundance and efficiency of the stomata are interesting points. In the typical land leaf the stomata occur as a rule more abundantly or exclusively on the lower surface of the leaf, but they occur only on the upper surface of the floating leaves of aquatic plants. They occur from 10 to 1,300 per square millimetre, but the usual number is round about 250, as in the apple and plum leaves. The stomatal pore is a narrow tube of elliptical cross section. In the sunflower the cross sectional area is about ·0001 sq. mm. when fully open, and as the stomata occur about 300 per sq. mm. the stomatal opening is about 3 per cent. of the total leaf surface.[1] Thus the immense numbers compensate largely for the small size of the stomata, and over and above this it has been ascertained that they are far more efficient in permitting gas diffusion than their cross sectional area would suggest. Thus the stomata of the sunflower epidermis permit diffusion of carbon dioxide into the leaf at a rate not 3 per cent. of the rate at which it would go into an open absorbing surface (such as an area of caustic soda solution equal to the total surface of the leaf[1]), but at so high a rate as 30 to 50 per cent. This is due to the additional diffusion which always goes on round the margin of pores. The demonstration of such great efficiency in the stomata has permitted physiologists to accept the view that these openings are indeed the main

[1] Since both sides of the sunflower leaf bear stomata, the leaf surface referred to here is twice the leaf area.

channels by which all gas exchanges of oxygen, carbon dioxide and water vapour, go on between the leaf and the outside air. Experimental evidence now leaves little doubt that this is so. It may be demonstrated in a simple way by an experiment with cobalt chloride paper. This paper when dry is bright blue, and it changes to pink as it absorbs moisture. If thoroughly dry pieces of the paper are placed on each side of leaves such as those of ivy or cherry laurel, the paper on the upper leaf surface will stay blue, and that on the lower surface will rapidly turn pink. This is due to the transpiration. which takes place solely from the lower surface in which the stomata occur. In leaves such as those of the water plantain (*Alisma plantago*) which have almost equal numbers of stomata on both leaf surfaces, the cobalt paper would turn pink on both sides at about the same rate.

Modern work on the stomata and the modern views about them well illustrate the present tendency of all physiologists to concern themselves with what a structure does, and the mechanism by which it takes effect, rather than to leave the situation with an unsatisfactory guess that the structure is present "in order to carry out some process" or "for some particular purpose" or "to serve some particular function." These latter can never be more than mere guesses, for the plant cannot be said to create structures *purposefully* at all, and progress can best come by examining physiological phenomena in physico-chemical terms, and by determining what things organs actually *do* and *how* they do them. The nature of the origin and persistence of organs is a matter to be considered closely in studies of heredity and evolution, and not to be guessed at prematurely.

CHAPTER XVII

THE FLOWER AND THE SEED

This is not the place to consider in detail the botanical complexities of floral structure and mechanism: we shall aim instead to present a sufficient outline of structure to shew something of the *principles* which underlie the reproduction of the following plant.

Although there is very great variety in the organisation of the floral parts, each flower on analysis appears to be a condensed shoot, bearing in spirals, or more commonly, in successive rings upon it, the specialised floral organs. The shoot never grows on again through the flower. The floral organs are said to be of two kinds, the *essential* and *inessential*, but naturally they are all essential in the sense that they combine to form one working mechanism. The "inessential" organs are the two outermost rings (or *whorls*). The outermost is the so-called *calyx* which is made up of *sepals*, and the innermost is the *corolla* made up of *petals*. It is usual in most flowers to find that the calyx is green, resembling a foliage leaf more than do the other floral organs, and enclosing and protecting the flower when it is in bud. The petals, by contrast, are very frequently brightly coloured, large and conspicuous, in relation to the attraction of pollinating insects. It should be recalled, however, that in one rather large section of the flowering plants there is no distinction between calyx and corolla, and these are represented by a single whorl of coloured structures, the so-called *perianth*. It should also be recognised that in many plants which are not insect pollinated the perianth or corolla is not necessarily large or conspicuous. The units which make up calyx, corolla, and perianth may either be separate, as for instance in the buttercup or lily, or joined to form tubular structures, as in the primrose or crocus.

The essential organs are in two series, the *androecium* and *gynaecium*, of which the former is outermost, i.e. lower on the floral axis. The units which make up the androecium are the so-called *stamens*, each of which has a relatively slender stalk or *filament*, and a thicker terminal part, the *anther*, in which the pollen is produced. The stamens are usually free from one another, but may sometimes be joined as in plants of the daisy family, and their filaments are frequently joined to the corolla. The gynaecium, which occupies the central part of the flower, consists of a box-like structure, the *ovary*, in which the ovules (which later become seeds) are produced. Surmounting the ovary is a stalk-like portion of very varied length, the *style*, which ends in the *stigma*. The stigma is an expanded, rough or papillate surface which is usually sticky, and which serves to receive the pollen grains. It frequently happens that the gynaecium is made up of more than one unit, so that numerous separate ovaries are found over the apex of the *receptacle*, the floral axis on which the flower parts are borne (see Fig. 75).

We have just described the commonest type of flower, in which both androecium and gynaecium are present together, but in some plants there are separate ovulate and staminate flowers, and in some, such as the willow (salix) one individual may bear only flowers of one kind. In this instance it is said that there are male and female trees, since the stamens produce pollen grains which form male gametes in the one, and the ovules produce female gametes in the other.

The Ovary. The structure of the gynaecium can best be seen in a simple example such as an ovary of the aquilegia. Here each unit in form resembles a simple foliage leaf folded round to meet at the margins. On this margin, and at the end of small veins, are to be found small thickened regions,

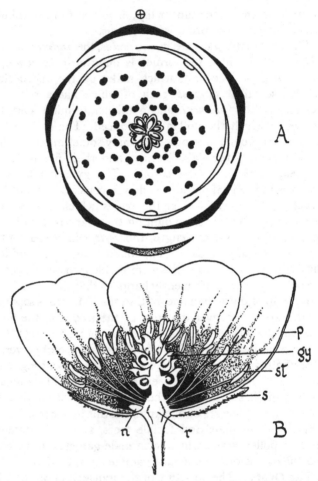

Fig. 75. *Ranunculus acris*, buttercup. *A*, the ground-plan, or *floral diagram*. *B*, the median elevation: *p*, petals, *s*, sepals, *st*, stamens, *gy*, gynaecium, *n*, nectary, *r*, receptacle. (After James and Clapham.)

the *placentas*, from which the ovules arise. Each such ovary is entirely closed, and terminated by a style and stigma: it produces five or ten ovules inside the central compartment. Each ovule arises by the development of an actively growing hump of tissue on the surface of a placenta. This hump enlarges to form an ovoid body, and as it grows two

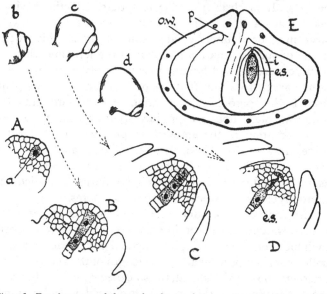

Fig. 76. Development of the ovule. *b*, *c*, and *d* shew stages in the growth of the young ovule as a hump on the placenta becoming gradually covered by the two integuments. *A—D* shew internal changes through this period, *a* is the archesporial cell, which divides with reduction-division to give a chain of four cells (*C*), of which three degenerate and one remains, the embryo-sac (*e.s.*). *E* is the ovary of larkspur in transverse section: *o.w.*, ovary wall, *p*, placenta, *i*, integument, *e.s.*, embryo-sac. (After Holman and Robbins, and Percival.)

collar-like rings of tissue grow up round it. These are the two *integuments* (see Fig. 76), and they extend until only a very small hole (the micropyle) is left at the apex. It very often happens, as in the figure, that the growth of the ovule stalk causes the ovule itself to take up an inverted position, with the micropyle pointing towards the placenta.

Within the developing ovule there soon appears one cell much larger than the others, this is the *archesporial* cell, and it can be shewn, in suitable microscopic preparations, to *divide by meiosis* to give four cells, of which only one, the *embryo-sac*, persists. It is clear from the fact that reduction-division takes place at this point, and also from a study of a very large number of intermediate plant types, that this process is one of spore formation like that in the fern, and that the embryo-sac is in fact homologous with a fern spore. It will be apparent from this, that the body of the flowering plant itself is the sporophyte generation. At the same time the flowering plant does not produce spores of one type only, for in addition to the embyro-sac produced in the ovules, there are also the *microspores*, i.e. pollen grains, formed in the stamens. There are some members of the Pteridophyta as well as the conifers and related forms, in which the sporophyte also produces two types of spores.

The Stamen. The chief interest of the stamen lies in the anther, which can be seen in cross section to be four lobed, each lobe containing a single elongated pollen-sac. The four pollen-sacs run parallel along the length of the anther, with a prolongation of the filament, the so-called *connective*, between them, carrying a rather weakly developed vascular stand. Each of these pollen-sacs resembles in development a fern sporangium, with which it is indeed homologous. As it approaches maturity, the pollen-sac is full of dense *pollen mother cells*, each of which *divides by meiosis* to give four pollen-grains, exactly as the spore mother cells give rise to spores in the fern. It is very interesting to find that the pollen-grains are liberated by a mechanism which distinctly resembles that which sets free the fern spores. The wall round each anther lobe is several cells thick, but at maturity it will be seen that the subepidermal layer is made of very regular large cells, the lateral walls of which are strengthened by conspicuous vertical bands of cellulose (see Fig. 77). This layer is incomplete in a strip on each side of the anther where the lateral lobes meet, and it is here that the opening begins. As in the fern, opening is brought about by drying, and an unequal contraction of the walls of the thickened

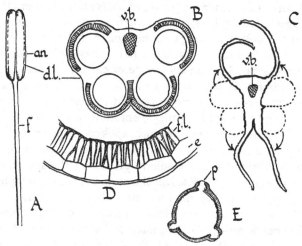

Fig. 77. *The stamen.* *A*, *an*, anther, *f*, filament, and *d.l.*, dehiscence line. *B*, transverse section through an anther shewing the four pollen-sacs, and round each the fibrous layer (*f.l.*); *v.b.*, the vascular bundle in the connective. *C*, the opening of the pollen-sacs by curling back of the wall at the dehiscence line and shrivelling of tissue between the lateral pairs of sacs. *D*, wall of the ripe anther shewing epidermis (*e*) and fibrous layer (*f.l.*). *E*, pollen-grain of birch with three projecting germ-pores.

cells. Contraction, following diminution of cell volume, can only take place in a tangential direction, and shrinkage of the wall, beginning at the lateral slits between the lobes, causes the anther walls to curl right back and expose the pollen-grains (Fig. 77, *C*).

As in the fern, the last stage of development of the spores (i.e. pollen-grains) is their development at the expense of other tissues within the spore-sac.

Pollen-grain. The pollen-grains are arranged tetrahedrally as a result of their origin by reduction-division, and they often retain signs of this in their shape or markings. The grains of insect-pollinated plants are usually covered with an outer membrane (*exine*) which is sculptured or pitted or covered with spines; these modifications are presumed to assist the pollen to cling to insects or to the stigma. Wind-pollinated plants, on the other hand, usually have smooth pollen-grains. The exine is made of cuticular material, but its continuity is broken by a number of *pores*, often one or three, but sometimes more, where the exine is thin, or is absent and exposes the inner membrane, the *intine*.

The Gametophytes. We have spoken of the embryo-sac on the one hand, and the pollen-grain on the other, as the final products of the sporophyte plants, i.e. the spores, and it is now necessary to discover the nature of the gametophyte structures which they produce when they "germinate."

The embryo-sac, remaining within the young ovule, undergoes a series of three nuclear divisions, so that it is occupied by eight nuclei. Of these, four are at each end of the much enlarged embryo-sac. Two nuclei, one from each end, now move towards one another, and fuse together in the centre of the cell to give the so-called *secondary nucleus*. The three remaining cells at the end away from the micropyle develop cell-walls: they are called the *antipodal cells* and play no further part in development. Of the three cells near the micropyle the most conspicuous is the "egg cell," the female gamete, and the other two cells are termed *synergidae*, since they are supposed, on rather slender grounds, to assist fertilisation. The embryo-sac at this stage represents the fully developed female gametophyte (Fig. 78).

Small as it is, that produced by the pollen-grain is smaller. When placed on a solution of cane-sugar, or upon the sticky surface of a stigma, the pollen-grain germinates by the rupture of one of the pores, and the protrusion through it of the intine, which forms a long slender tube resembling a fungal hypha. Meanwhile there have been nuclear changes within the cell. Before the grain is shed its nucleus is divided into two parts, called respectively the *generative cell* and the *tube nucleus*. After germination the generative cell moves into the apex of the pollen-tube and there divides to form

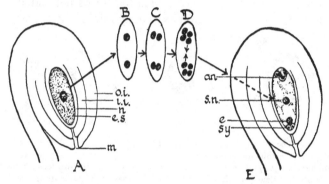

Fig. 78. Development of the female gametophyte. *A*, ovule with outer integument (*o.i.*), inner integument (*i.i.*), nucellus (*n*), and embryo-sac (*e.s.*); *m*, micropyle. *B*, *C* and *D* shew nuclear divisions within the embryo-sac; in *D* the arrows shew the approach of two nuclei which will fuse to give the secondary nucleus (*s.n.* in *E*). *E*, the ovule ready for fertilisation; within the embryo-sac are seven nuclei, three in the antipodal cells (*an*), the secondary nucleus (*s.n.*), the egg cell (*e*), and two synergidae (*sy*), near the micropyle.

two male nuclei (Fig. 79). Thus the male gametophyte consists of no more than an envelope containing three nuclei, of which two are male gametes. Small as these gametophytes are, their origin and behaviour leave no doubt that they are an equivalent stage in the life-history of the flowering plant to the green prothallus in that of the fern.

Pollination and Seed Formation. Pollination is the transfer of pollen-grains from the stamen to the stigma of the same or another plant. In the vast bulk of flowering plants this transference is made by insects, and it seems clear that the colour of the floral organs, the patterns which they form, their shapes, their arrangement into heads, their scents, and their production of nectar in special secretory structures, are all devices which assist in the attraction of insects, and facilitate this pollen transference. In many instances particular flowers are clearly specialised to the visits of particular insects and can hardly be pollinated by others. There is a large and extremely interesting literature

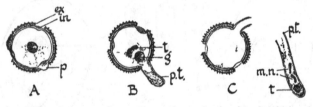

Fig. 79. The male gametophyte. *A*, a pollen-grain shewing the thick exine (*ex*), and the thin intine (*in*) exposed at the three pores (*p*). *B*, protrusion of pollen-tube (*p.t.*): the nucleus has divided into the tube-nucleus (*t*), and generative nucleus (*g*). *C*, at full maturity the apex of the pollen-tube contains the tube-nucleus and the two male nuclei (*m.n.*) formed from the generative nucleus. (After Bonnier and Sablon.)

on the subject of such relationships, and more particularly on the arrangements by which cross-pollination (i.e. pollination of one plant by pollen from another) is encouraged, and self-pollination avoided.

It is interesting to examine the flower of the buttercup (*Ranunculus*) as a typical, not very complicated example of pollination mechanism (Fig. 75). Colour attraction is chiefly with the five glossy yellow petals, which have rather faint radial markings extending from their bases, and with the large mass of numerous free yellow stamens. At the base of each petal is a nectary covered by a small scale which is attached basally, so that it forms a pocket to which nectar gathers. This nectar is a primary attraction to insects, but many also take the pollen itself. The insect visitors are short-tongued flies and bees, including the hive-bee. When the flowers open the stamens are erect and packed round the gynaecium, so that there is a space between them and the petals by which insects can reach the nectaries. The stamens progressively dehisce, the outermost first, and as they shed their pollen the filaments bend outwards, so that the empty anthers lie upon the petals. In this way there is always a path to the nectary with dehiscing stamens along it, and insects reaching the nectar become covered with the pollen. The narrow stigma on each of the many free carpels is apparently receptive from the opening of the flower, but it is when the last stamens have dehisced and bent away from the gynaecium that pollination is most likely. There is thus some small effect in favour of cross-pollination, but self-pollination is quite feasible. The flower is open and simple in structure, and not specialised to the visits of one type of insect alone. At the same time, the position of the nectaries and the stamen movements shew a fairly high degree of organisation.

A very large group of plants, including the grasses, sedges and rushes, on the one hand, and many trees on the other, are habitually wind-pollinated, and their pollen is freely distributed, although, of course, it does not long retain its viability. Other, much smaller groups of plants, are specialised to pollination by free water and even by birds (e.g. the *Fuchsia*, pollinated by sun-birds).

Whatever the method of transference to the stigma, once there, the pollen-grain germinates and sends out a pollen-tube which grows into the loose tissues of the stigma, towards the ovary. Sometimes it grows along a small stylar passage, but often it traverses tissues directly, breaking them down just like a fungal hypha. Its growth is evidently directed by a secretion from the ovary, for if a small portion of ovary is placed at the edge of a sugar solution in which pollen is germinating, the pollen-tubes will all be directed towards it.

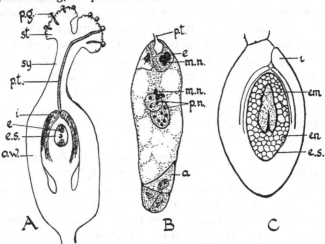

Fig. 80. *Fertilisation.* *A*, longitudinal section through the gynaecium of *Polygonum*, shewing pollen-grains (*p.g.*) germinating on the stigma (*st*), and sending pollen-tubes (*p.t.*) through the style (*sy*) towards the embryo-sac (*e.s.*); *i*, integuments; *e*, egg cell; *o.w.*, ovary wall. *B*, the embryo-sac at fertilisation, shewing the two male nuclei (*m.n.*) fusing, one with the egg cell (*e*), and the other with two polar nuclei (*p.n.*) which have not yet themselves fused; *a*, antipodal cells; *p.t.*, pollen-tube. *C.* longitudinal section through an ovule some time after fertilisation. Within the embryo-sac (*e.s.*) the fertilised egg cell has given rise to the embryo (*em*), and the nucleus resulting from fusion of two polar nuclei and a male gamete has divided to form the nutritive endosperm tissue (*en*). (After Strasburger and Guignard.)

The pollen-tube crosses the air-space in the ovary, or grows through the ovule stalk until its apex enters the micropyle (Fig. 80). The time taken for the pollen-tube to grow from the stigma to the micropyle is extremely variable, ranging from a few hours in some plants, to several months in others.

Within the micropyle, the tip of the pollen-tube opens and the two male nuclei are liberated into the embryo-sac. One of them fuses with the egg cell, to produce the zygote from which the new embryo arises, and the other fuses with the secondary nucleus. This process is the act of "fertilisation." The product of fusion with the secondary nucleus has by this time involved three nuclei, and will eventually form a food reserve for the young embryo, the *endosperm*.

The fertilised egg cell within the embryo-sac now commences a series of divisions which will give rise to the

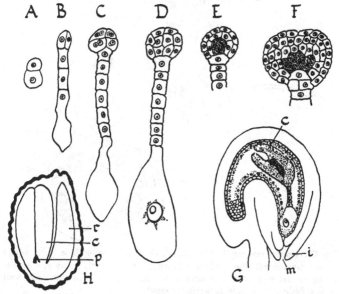

Fig. 81. *Formation of embryo.* *A—H* form a series shewing the divisions of the fertilised egg giving rise to an embryo in Shepherd's purse (*Capsella bursa-pastoris*). In *G* the integuments (*i*) and micropyle (*m*) are shewn; the cotyledons (*c*) are recognisable, and the embryo is surrounded by endosperm. *H* shews the ripe seed filled with the embryo which clearly shews cotyledons (*c*), plumule (*p*) and radicle (*r*) inside the thick testa which has formed from the integuments. (After Coulter and Chamberlain, Bergen and Caldwell, Fritsch and Salisbury.)

embryo plant (sporophyte) within the seed. These divisions
are seen in Fig. 81. The divisions form a short row of cells,
at the micropylar end of which is a single large basal cell,
whilst at the other is a head which undergoes repeated
divisions to form a structure, at first heart-shaped and then
bilobed. In most plants the embryo is now quite visible, and
the two lobes are recognisable as the "*cotyledons*" or seed-
leaves, whilst between them is the rudiment of the young
shoot, the "*plumule.*" The conical base of the embryo is the

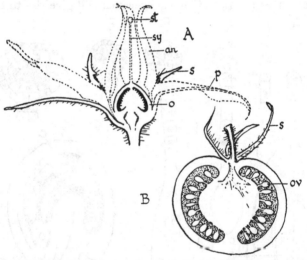

Fig. 82. Fruit formation in tomato. *A*, the flower, with those parts which
persist drawn in solid line, and those which are shed, dotted. *o*, ovary, *s*, sepals,
st, stigma, *sy*, style, *an*, androecium, *p*, petals. *B*, the fruit with ovary wall
become fleshy: *ov*, ovules, *s*, persistent sepals.

young root, or "*radicle.*" During this development the
embryo grows by food supplies from the parent plant, and
in some plants, at the expense of the endosperm tissue
formed around it by division of the fertilised secondary
nucleus. In other plants, such as the castor-oil (*Ricinus*),

the endosperm remains in large amount within the seed beyond the time of germination. The effects of fertilisation extend much beyond the embryo-sac itself, the ovule enlarging and maturing, to become a "seed" and the ovary to become the fruit (Fig. 82). The ovary wall may be extremely variously modified to form structures relating to many methods of seed-dispersal. It is also common knowledge that fertilisation leads, as a rule, to the withering and dropping of other parts of the flower, such as the stamens, corolla and calyx. The modified integuments persist as the so-called "testa" or seed-coat, and this also may have sculpturings, hairs, plumes or wings formed from it, which assist dispersal. As with pollination mechanisms, so with the dispersal of the ripe seed, there is very wide and interesting adaptation to dispersal by various agents, particularly the wind, different kinds of animals, and mechanical forces liberated by drying and splitting. It is not proposed to give examples of these mechanisms, which are familiar enough to everyone, but it may be pointed out what an extraordinarily close adaptation they shew between the flowering plant and its terrestrial environment.

As a rule the water content of the ripe seed is very low, and its vital activities appear to be almost suspended; for instance the respiration rate is so low as to be hardly detectable. Either in the cotyledons or in the endosperm the seed contains abundant food reserves, a fact which is the direct cause of the high food value of so many seeds (e.g. cereals, legumes) to the human race. The seed in this state, moreover, is extremely resistant to changes in the external environment, and so long as it remains dry will withstand a considerable time of immersion either in boiling water or in liquid air. As a resistant resting structure it represents the highest organisation in the plant kingdom.

Germination. If certain conditions are satisfied, the seed after dispersal, will germinate. The first of these conditions is the viability of the seed itself. This is impaired by many kinds of adverse conditions, but more particularly by increasing age. The longevity varies very greatly, from a very few days for the small seeds of willow and poplar (which must not in any case have dried) to legumes with thick testas, which have germinated when taken from authentic museum collections over fifty years old. There is some evidence that water-lily seeds may sometimes remain viable for as long as two hundred years, but stories of the germination of "mummy wheat" are quite unproven, and the plants grown from purchased packets of "seeds from mummy cases," oddly enough, have often turned out to be American species alien to Africa. A second general condition for germination is the presence of water, which plays many parts in the germination process. For example, (1) it softens the seed-coat, (2) it is involved in imbibition and bursting of the testa, (3) it makes the testa more permeable to gases, facilitating the entry of oxygen and the escape of carbon-dioxide. In all probability it plays a part in the increasing activity of the protoplasm, in the hydrolysis of reserve materials by enzymes, and in the translocation of soluble organic matter to actively growing parts. Not only must the seed be viable, but it is further necessary for germination that the temperature should lie within a certain favourable range, which, however, is not quite the same for all plants. As might be expected from the vastly increased respiratory activity of the germinating seed, oxygen is also necessary, and it can easily be seen that seeds kept in boiled water will fail to germinate though other conditions are favourable. In addition to these conditions, germination is affected by light or darkness, but different species behave differently in relation to this factor. The seeds of *Lythrum salicaria*, the purple loosestrife, appear to need light for germination, whilst the seeds of many other plants will germinate only in complete darkness. It will be realised that the need for light in the *germination* of some seeds is an entirely different matter from the need for light for photosynthesis in the *establishment* of the young autotrophic plant. The germination of seeds of many wild plants is further complicated by the condition of *dormancy* which they can assume. In such a state the seeds fail to germinate although placed under conditions which at other times will cause them to germinate. The causes of such dormancy are very numerous, but the advantage to the plant is clear in that it makes for a delayed germination, and the production of new seedlings over a relatively long period. During the Great War of 1914–18 many old pastures were ploughed up for wheat growing. The new fields at once became a blaze of charlock plants which could only have grown from seeds buried and dormant in the soil since the time, many years before, when the pastures were laid down from arable land.

The external evidence of germination is the splitting of the testa at the micropyle, and the growing out of the radicle, which at once penetrates the soil and becomes anchored there by the production of root-hairs. By the time this has happened, active extension has begun in some other part of the embryo plant. If this is in the *hypocotyl* (that is the region between the radicle and the cotyledons) the cotyledons and the plumule, still in the seed-case, will be carried above ground, and there, at a later stage, the cotyledons as well expand and the plumule will extend (Fig. 83, *A*). This type of germination is found in sunflower, marrow, castor-oil, etc. If the extension is in the *cotyledons*, these structures elongate, and their apex carries the testa up into the air, the plumule growing out afterwards from just above the top of the radicle. The onion (with one cotyledon only) shews this. If the *plumule* extends at once, then the cotyledons

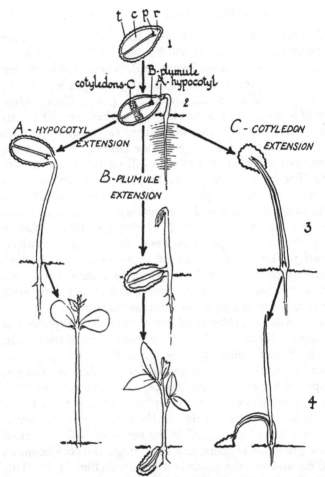

Fig. 83. Germination types in relation to the region of the embryo in which extension next proceeds after the radicle has once become firm in the soil. 1 shews the seed with testa (*t*), cotyledons (*c*), plumule (*p*) and radicle (*r*). In 2 the radicle is established and there are shewn dotted the three zones in which extension may next go on. In hypocotyl extension the cotyledons and testa come above ground and the plumule only later grows from between them. In plumule extension the cotyledons remain at soil-level. In cotyledon extension although the testa is carried into the air the plumule remains at soil-level, and then grows upwards from this level.

will be left undisturbed on or below the soil (e.g. pea or acorn). The effect in all these instances is the utilisation of the stored food materials and the establishment of a new plant. In some seedlings the cotyledons themselves become green and behave as ordinary photosynthetic foliage leaves (castor-oil, marrow), but in others, such as the pea, the cotyledons are purely storage organs.

Conclusion. From what we have written in this extremely brief survey, it will be clear that the flower and seed production represent tremendous advance from the fern in solving the problem of sexual reproduction independently of liquid water. Microspore dispersal (i.e. dispersal of pollen) is now dependent not only on the wind, but upon insects, and the evolution of the flower can be shewn to have gone parallel with the evolution of insects. The dispersal of the macrospore (embryo-sac), on the other hand, is so much delayed that it germinates in the protection of the parent plant, produces the gametophyte and its gametes there, is fertilized there, and develops a new sporophyte there, which also receives nourishment and protection from the parent sporophyte plant. The structure which is thus shed, the seed, has an extremely elaborate organisation, consisting, as it does, of a young sporophyte within the remains of an old gametophyte, enclosed again in the integuments of the parent sporophyte. The dispersal of this structure is achieved by very complex and specialised mechanisms. Even in its physiological responses that relate to germination, the seed has reactions which favour establishment.

No less than the vegetative structure and mechanism, flower and seed production represent the height of specialisation of plants to conditions of life on the land.

INDEX

Chief references in italics

Printed in the United States
By Bookmasters